The Chemistry of God

ELOHIM

Exploring the chemistry laboratory from A to Z
Uncovering the profound Biblical significance within

Dr. Laurenee London Adeoshun, Ph.D.

Laurenee London, Ph.D.

ISBN: 979-8-89705-796-2

At **thechemistryofgodseries.com**, you'll discover *The Chemistry of God* educational series, including resources like the **The Chemistry of God:** Exploring the Chemistry Lab..., **General Chemistry Workbook**, **Chemistry Adventures with Dr. London, Guess The Elements of God** interactive game, and more. This collection is designed to bridge scientific understanding with faith, offering engaging tools for learning and spiritual growth.

We encourage you to share these transformative resources with educational institutions to spread the truth of Jesus Christ and advance the Kingdom of God. You can also request this captivating series at your local libraries to inspire curious minds seeking both knowledge and faith.

Additionally, we value your thoughts! On our website, you have the opportunity to review and share your feedback about the series. Your insights will help us continue to refine and expand this work for the glory of God.

At *The Chemistry of God Ministries*, we deeply value you— but remember, Jesus Christ loves you even more! Stay blessed, and remain a faithful and radiant bride of Christ.

TO ELOHIM

Heavenly Father, I am deeply grateful for all You have done to align my life with the vision You have for me. Thank You, Elohim, for my husband, Raphael, our children, and our extended family members. Thank You for being at the center of our household, marriage, businesses, and relationships. Let Your pleasing will continue to be done in our lives on earth as it is in the Kingdom of Heaven, according to Matthew 6:10.

Thank You, Lord, for guiding me through each trial and every moment of hardship, which has strengthened my relationship with You and prepared me to fulfill this calling. I am especially grateful to those who, perhaps unknowingly, brought turmoil and challenges into my life. Through their actions, You sharpened my character and prepared me to pass on the wisdom and understanding I've gained.

Lord, I am in awe of Your divine plan. I recognize that nothing happens without Your knowledge and that everything serves a purpose under Your sovereignty. I pray our relationship continues to deepen and expand, far beyond what I can now comprehend. Most importantly, Lord, I hope my actions have brought You joy and honor. Thank You for being my Father, my guide, my closest companion, and the unfailing love of my soul. I love You deeply and rejoice in knowing You are alive, present, and fully in control.

FOR THE FRUITS OF MY WOMB

This book is dedicated to my children, the fruits of my womb.

Thank you, Lord, for blessing me with wise and Holy Spirit-filled Sons of God. Father, I dedicate them back to You to be used for Your glory, adoration, and praise. I am grateful that You have blessed the work of their hands and ordered their steps, in keeping with Your word in Psalm 37:23: *"The steps of a good man are ordered by the Lord, and He delights in his way."* Thank You, Jesus, for surrounding them with Your spirit of favor as with a shield and for granting them a spirit of ease.

I also declare Psalm 1:1-3 over their lives: *"Blessed is the man who walks not in the counsel of the ungodly, nor stands in the path of sinners, nor sits in the seat of the scornful; but his delight is in the law of the Lord, and in His law he meditates day and night. He shall be like a tree planted by the rivers of water, that brings forth its fruit in its season, whose leaf also shall not wither; and whatever he does shall prosper."*

Sons, I pray that you continue to thrive in your relationship with the Lord, in your education, your businesses, and in all the kingdom assignments He entrusts to you. May you accomplish even more than I have through the strength of the Lord. Remember always, I love you deeply, but Jesus Christ loves you even more.

Contents

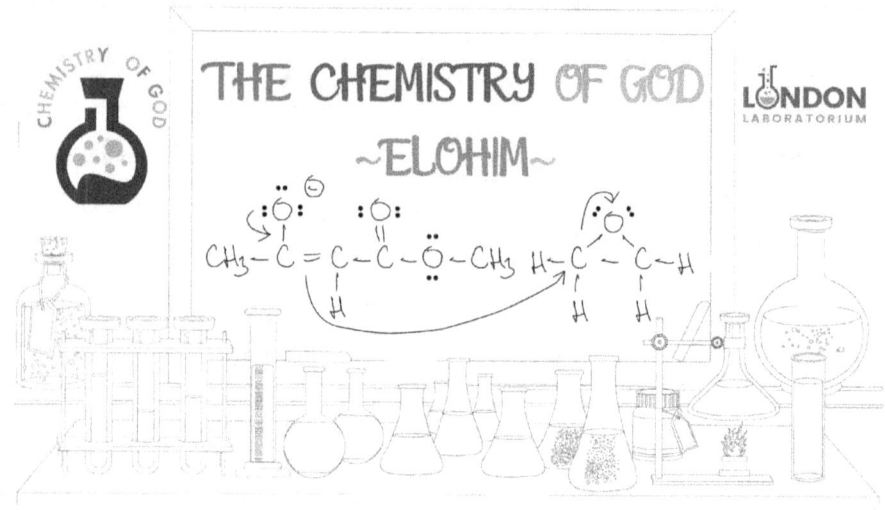

Introduction

"The Chemistry of God" invites readers on a captivating journey through the intersections of science and the spiritual realm of God. Across its chapters, the book explores various laboratory instruments and procedures, revealing the profound moral insights they hold. By weaving together the wonders of science with the timeless wisdom of the Bible, this book illuminates the deep connections between these seemingly distinct worlds.

In keeping with the practices of *The Chemistry of God Ministry*, we honor Elohim, Jesus Christ, and Holy Spirit, —the Holy Trinity— by consistently capitalizing their names and pronouns as a sign of reverence. Conversely, we deliberately use lowercase letters when referring to satan and his followers, to withhold any recognition or honor. Moreover, we begin with the letter "Z" instead of "A" as a symbol of our Zeal for the Lord.

Prayer of Consecration

Heavenly Father, in the mighty name of Jesus Christ, I call upon Your holy and righteous name to ask for holiness, righteousness, and purity.

In the living name of Jesus, I decree that I am found blameless, righteous, and upright like Your servant Job. I purge my foundation and bloodline of iniquity, rebellion, rejection, and bitterness by the power in the blood of Jesus Christ of Nazareth.

Father, I decree, in the mighty name of Jesus, that my heavenly garments are restored, spotless, and blemish-free now by Your cleansing blood.

Lord, I stand on the promises of Your living word, which says in Psalm 51:7, *"Purge me with hyssop and I shall be clean; wash me and I shall be whiter than snow,"* and so shall it be now in the mighty name of Jesus.

Jesus, I decree that Your Dunamis resurrection power mends and purifies my mind, body, soul, spirit, and heart, for You say, "Be holy, for I am holy," in the name of Jesus.

Father, I cannot do life alone. I need Your help, Holy Spirit, to guide my footsteps and help me keep them planted firmly on the fine and narrow path to my Lord and Savior, Jesus Christ of Nazareth.

Jesus, help me to make You my Lord in every area of my life as I surrender all of myself to You. Thank You, Jesus, for Your living water that pours out in Isaiah 61:3, saying, *"To console those who mourn in Zion, to give them beauty for ashes, the oil of joy for mourning, the garment of praise for the spirit of heaviness; that they may be called trees of righteousness, the planting of the Lord, that He may be glorified."* I receive them now in Jesus' mighty name.

Furthermore, I decree Isaiah 61:10: *"I will greatly rejoice in the Lord, my soul shall be joyful in my God, for He has clothed me with the garments of salvation and covered me with the robe of righteousness as a bridegroom or bride decks themselves with ornaments or jewels."* In the mighty name of Jesus Christ, I pray, and so shall it be. Amen.

Figure 1: A Dove with an olive branch is a symbol of peace, love, and reconciliation to Elohim.

Zeal for the Lord – Authentic Worship

The concept of "zeal for the Lord" emphasizes the fervent and sincere devotion that God expects from His people. To have a burning desire to please God, to do His will, and to advocate His glory in the world in every possible way is to have the zeal of the Lord. This devotion is rooted in the biblical principle highlighted in John 4:24: *"God is spirit, and his worshipers must worship in Spirit and in truth."* The expression "zeal for the Lord" reflects a passionate and wholehearted commitment to God. It involves a genuine and authentic worship that goes beyond outward works and ceremonies. God desires worship that emanates from the innermost being of individuals (our soul and heart), a worship driven by a deep connection with the Spirit of God.

The call to worship in Spirit and truth implies a pure and honest engagement with God. It invites believers to approach their relationship with God with sincerity, authenticity, and a genuine passion for His presence. This concept encourages a heartfelt devotion that transcends mere religious practices, emphasizing the importance of a personal and intimate connection with Jesus Christ. In essence, "zeal for the Lord" reminds believers to develop an intimate relationship with Jesus Christ, to live out one's faith with passion, authenticity, and a deep connection to the spiritual reality of God.

Zeal for the Lord – Love Disarms

In our own strength we can only go so far, but when we tap into the supplier of our strength, Jesus, we regain momentum and are able to continue our journey of vision. As children of God, we are the target of the devil because he is bitter, jealous and envious of our status in Christ Jesus. The devil and his cohorts have already been judged and, at the appointed time, they will perish in the lake of fire in hell. Knowing this, the enemy seeks to take down as many people as he can to spite Jesus, aiming to have as few believers as possible when our Lord and Savior returns to earth for the rapture to take place. Psalm 34:19 encourages believers; *"Many are the afflictions of the righteous, but the Lord delivers him out of them **all**. He guards **all** his bones; not one of them is broken. Evil shall slay the wicked, and those who hate the righteous shall be condemned. The Lord redeems the soul of His* servants, and none of those who trust in Him shall be condemned." Further reassurance of God's protection is found in the promise of Luke 10:19; "Behold, I give unto you power to tread on serpents and scorpions, and over **all** the power of the enemy: and nothing shall by any means hurt you." Don't worry about anything, surrender every concern to the Lord and He shall surely deliver you.

The Sons of God are in constant battle against the enemy and often feel the burdens of the world more than others because the devil attacks those who belong to God, not those he already holds captive. Keep in mind that Jesus and the disciples faced countless trials and tribulations throughout their ministry. John 15:18-20 says, "If the world hates you, keep in mind that it hated me first. If you belonged to the world, it would love you as its own. As it is, you do not belong to the world, but I have chosen you out of the world. That is why the world hates you. Remember what I told you: 'A servant is not greater than his master.' If they persecuted me, they will persecute you also. If they obeyed my teaching, they will obey yours also."

The inevitable obstacles and challenges we face in life—whether financial hardships, health issues, relationship struggles, or spiritual warfare—can leave us feeling overwhelmed and drained. However, when we renew our strength in the Lord daily through prayer, praise, and worship, we tap into an everlasting source of power that sustains us through the trials of the world. These trials aim to increase our faith and train us to enter and thrive in our promised land. They test us to apply what we have been taught in the Lord, with the answer for every test being Jesus. When we surrender our worries to the Lord, we demonstrate our faith that He is able to win the battle, rather than placing our trust in man or false gods. Remember, God is the "I Am," capable of providing all our needs. The Lord recommends that we pray effectively by using His name that best fits our current area of need.

Name of God	Meaning	Name of God	Meaning
Abba	Father	Alpha and Omega	The Beginning and End
Adonai	Lord and Master	Jehovah Rapha	The Lord That Heals
Elohim	God | Judge | Creator	Jehovah Shalom	The Lord Is Peace
El Shaddai	All Sufficient One Lord God Almighty	Jehovah Jireh	The Lord Will Provide
El Elyon	The Most High God	Jehovah Raah	The Lord My Shephard
Yahweh	Lord | Jehovah	Jehovah Sabaoth	The Lord of Hosts
El Olam	The Everlasting God	Jehovah Mekadesh	The Lord Who Sanctifies You
El Roi	The God Who Sees	Immanuel	God With Us
El Gibor	The Mighty God	Jehovah Hoseenu	The Lord Our Maker
Qanna	Jealous	Jehovah Nissi	The Lord My Banner
Jehovah Tsidkenu	The Lord Our Righteousness	Jehovah Shammah	The Lord Is There

Figure 2: Names of God and their divine meaning.

It's important to note that not every trial is from the Lord; some tests come from the devil and are usually distractions meant to steal the incoming blessings of the Lord in our lives. The enemy is cunning and will bring trials to cause us to sin, gaining the legal right to divert our blessings to the kingdom of darkness. The devil knows when the children of God are scheduled to be rewarded or promoted and will lay traps and obstacles to ensnare them. This is why we need to be in constant prayer to withstand the tricks of the devil. When we maintain a heart posture of praise and purity, we are able to receive the blessings of the Lord. But the enemy seeks to stain our garments with sin, so we are no longer in a pure state to receive these blessings, thus stealing them.

However, when we are fed up with the devil's tactics, we can humble ourselves before the Lord through fasting, prayer, and repentance and the Lord will restore us. This promise is in the second book of the Chronicles 7:14; *"If My people who are called by My name will humble themselves, and pray and seek My face, and turn from their wicked ways, then I will hear from heaven, and will forgive their sin and heal their land."* Joel 2:25 reassures the promise; *"And I will restore to you the years that the locust hath eaten, the cankerworm, and the caterpillar, and the palmerworm, my great army which I sent among you."*

The constant battle between good and evil can be tiring, so when we feel weary, we must encourage ourselves in the Lord and reconnect with our source of strength, Jesus Christ of Nazareth. Matthew 11:28 says, *"Come to me, all you who are weary and burdened, and I will give you rest,"* demonstrating that the Lord will provide peace, comfort, and strength when we surrender our cares and worries to Him. Isaiah 41:10 promises, *"Fear not, for I am with you; be not dismayed, for I am your God. I will strengthen you, yes, I will help you, I will uphold you with My righteous right hand."*

As children of God, we must renew our strength daily and wait on the Lord in faith because His timing is best. Isaiah 40:31 contains a great promise of strength for the weary: *"But those who wait on the Lord shall renew their strength; they shall mount up with wings like eagles, they shall run and not be weary, they shall walk and not faint."* Though the vision or promises of the Lord may tarry, they will surely come to pass by the grace of God. When you feel down and exhausted, remember the popular scriptures for motivation to keep on the garment of praise and worship. Psalm 30:5 says, *"For His anger is but for a moment, His favor is for life; weeping may endure for a night, but joy comes in the morning,"* and Philippians 4:13 says, *"I can do all things through Jesus Christ who strengthens me."* Encourage yourself in the Lord, speak victory during the test, and you will overcome. Lastly, love is always the answer; Jesus is love, and love disarms the enemy of the children of God.

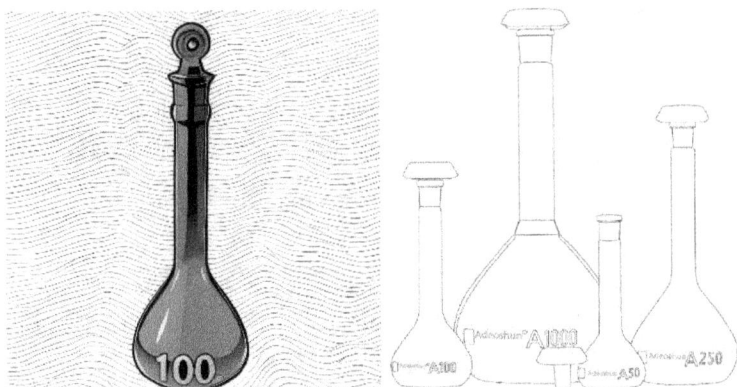

Figure 3: Actinic Volumetric Flasks are designed to limit a sample's exposure to light, by using amber or dark-colored glass that blocks UV and Visible light.

Actinic Glassware: A Special Flask

Actinic glassware is used for chemicals that are sensitive to light, protecting them from the bright and colorful light around us. A popular type of actinic glassware frequently used in the laboratory is the volumetric flask.

A volumetric flask is a container with a round bottom, a long neck, and a flat base. It's used by scientists to measure the exact amount of a liquid. The neck of the flask has a tiny line that indicates how much liquid to pour inside. When you look at the liquid in the flask, it forms a little curve or half-circle on the surface called the meniscus. This curve is a guide that helps us measure liquids accurately.

To use the volumetric flask properly, make sure the lowest point of the meniscus lines up with the marking on the neck. Keep the flask level to get the right volume reading.

Biblical significance: A Protective Shield

Just like actinic glassware protects chemicals, there's a special kind of protection from the Lord mentioned in the book of Job in the Bible. It discusses the most powerful being, Elohim (also known as God, Lord, or Father), forms a protective shield around His people and everything they own. It's a divine protective barrier around our home and belongings.

In Job 1:10, satan the adversary of the children of God says, *"Have you not put a hedge around him and his household and everything he has?"* This means the enemy has to seek permission from God to pursue His children and that everything good in a believer's life is guarded and shielded. Just like actinic glassware keeps the chemicals safe from destruction by light, Elohim keeps His people who are obedient to his commandments safe and secured. In Psalm 91, the Lord illustrates his ability to protect His children in time of trouble and having a strategic plan to command His angels to guard us in all ways, so we are not harmed. It's a personalized and supernatural kind of care, having the Most High living and powerful God looking out for us! Therefore, there is no need to fear satan and his tactics because, as believers and doers of the word of God, following His commandments, the enemy can't truly harm us. When you feel afraid, remember that the Lord did not give us the spirit of fear but of love, power and a sound mind to make wise decisions, as referenced in 2 Timothy 1:7. Reject the spirit of fear and trust in the divine protection and care of the Lord.

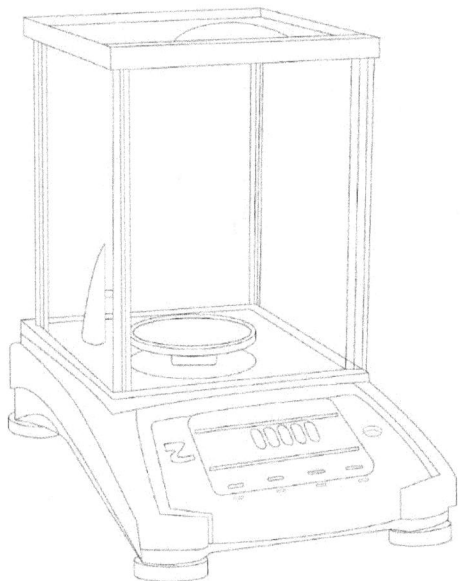

Figure 4: Analytical Balances are designed to measure small masses with a high degree of accuracy.

Balance – A Tool for Precision

A balance is a tool used to determine the weight of an object. When using a balance, it's crucial not to move it or press down on the surface it's sitting on. That's because balances are set up very carefully for the exact spot they're in. Think of it like a super-sensitive scale that needs to stay perfectly still to give an accurate weight measurement.

Biblical significance: Honest Weighing

Just like we must be careful not to disturb the balance for accurate measurements, there's a lesson in the Bible, in Proverbs 16:11. It says, *"Honest balances and scales are the Lord's; all the weights in the bag are of his making."* This teaches us that God wants us to use fair and truthful balances in our lives. In our everyday actions and business dealings, we should be like the careful balance, working with accuracy and integrity. This way, we are following God's guidance and making sure our actions are fair and just, just like the precise measurements on a balance. By doing this, we are held blameless and remain on the path to greatness in the Lord.

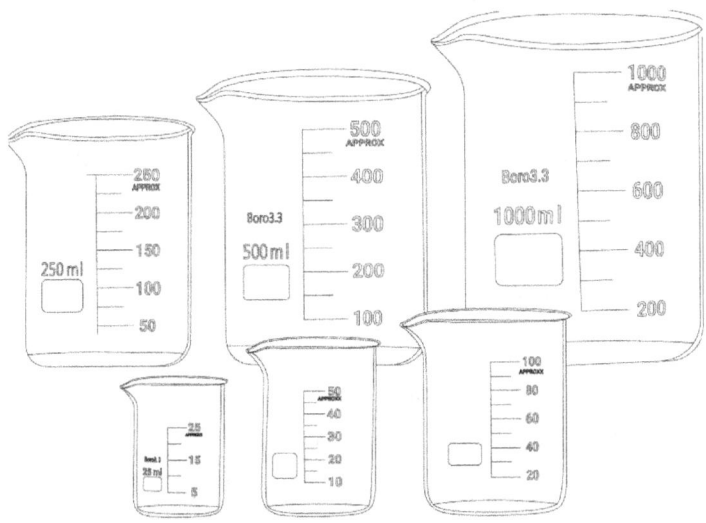

Figure 5: Beakers are versatile lab containers used for mixing, heating, and measuring liquids.

Beakers – Versatile Tools in the Lab

Beakers, a frequently used tool in the laboratory, are used for stirring, mixing, and heating chemicals. These useful containers have special markings that help scientists measure the amount of liquid they hold, and they come with convenient spouts on the rim for easy pouring.

Biblical significance: Honesty in Measurement

Just as beakers are used to measure substances accurately, there's a valuable lesson in the Bible about the importance of honesty, as found in Proverbs 11:1: *"The Lord detests dishonest scales, but accurate weights find favor with him."* This teaches us that God values honesty not only in the physical measurement of objects but also in our life decisions and interactions.

In Matthew 18:21-35, Peter asks Jesus how often he should forgive a brother who sins against him. Jesus' response indicated that we should forgive an innumerable number of times, illustrating this point with a parable. In the parable, the kingdom of heaven is likened to a king who decides to settle accounts with his servants. One servant owes the king a vast sum but cannot pay. When the servant pleads for more time, the king, moved by compassion, forgives the entire debt. However, this same servant later encounters a fellow servant who owes him a much smaller amount. Instead of showing mercy, he demands immediate payment and has the fellow servant thrown into prison when he cannot pay. Upon hearing of this, the king rebukes the servant, saying, *"You wicked servant! I forgave you all that debt because you begged me. Should you not also have had compassion on your fellow servant, just as I had pity on you?"* The king then delivers the servant to be punished until he can repay his debt.

This parable highlights the expectation that those who receive forgiveness must extend the same grace to others. It underscores the moral that, just as we seek forgiveness from God for our own shortcomings, we should also forgive those who wrong us. The story emphasizes that using unjust and dishonest measures in our dealings with others, like the servant who was unforgiving despite having been forgiven, goes against God's principles.

The lesson is clear: by being truthful, fair, and forgiving, we align ourselves with God's values and find favor in His eyes. This principle is echoed in Matthew 7:12: *"Do unto others as you would have them do unto you."* This Golden Rule serves as a divine recipe for a life of integrity and righteousness. By adhering to this principle, we build trust and goodwill, reflecting the fairness and justice that God desires from us.

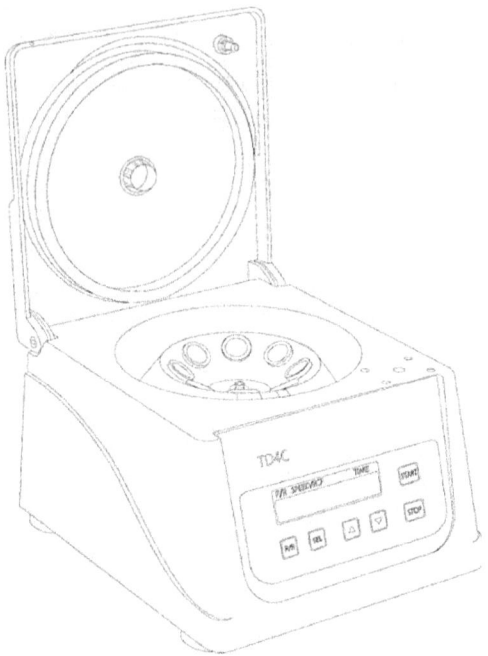

Figure 6: Centrifuges are devices used to separate substances based on particle size and density by spinning samples at high speeds.

Centrifuge – Spinning Separation Phenomenon

A centrifuge spins mixtures at high speeds to separate their components. The heavier or denser components settle at the bottom of a test tube, leaving a clear liquid, free of any solids, on top. It's like an instant sorting hat for liquids, quickly and efficiently organizing them based on their densities, which depend on the mass of the substance and the volume or space it occupies.

Biblical significance: Separation for Righteousness

Just as a centrifuge skillfully separates mixtures, there's a lesson in the Bible, found in Deuteronomy 18:9. Moses, who led the Israelites out of the hands of their oppressors in Egypt, says, *"When you enter the land the Lord your God is giving you, do not learn to imitate the detestable ways of the nations there."* Furthermore, in Deuteronomy 23:6, God instructs the Israelites to stay separate from other nations. Making peace treaties might compromise their relationship with God because they could be influenced to worship false gods. This teaches us that, like a centrifuge separating mixtures, God expects His people to distance themselves from practices that go against His ways.

The story of the Gibeonites deceiving the Israelites in Joshua 9:14-15 serves as a cautionary tale about the consequences of making decisions without seeking divine wisdom through prayer. Despite God's instruction to remain separate from other nations, the Israelites made an uninformed decision that led to unintended outcomes and disobedience to God's commandments.

The Gibeonites, from a neighboring land, devised a plan to save themselves from God's judgment by pretending to be foreigners. Their deception was elaborate, involving worn-out clothing, old wineskins, and molded bread to create the appearance of travelers from a distant land, so the Israelites would make a peace treaty with them. Their fear of being uprooted by the Israelites drove them to resort to deceitful tactics to secure their safety.

This incident highlights the broader context of God's promise to give the Israelites the land of Canaan due to the wickedness and rebellion of its inhabitants against God's commandments. The Gibeonites' actions illustrate the lengths to which people may go to avoid the consequences of their disobedience.

Despite discovering the deception of the Gibeonites, the Israelites chose to honor their agreement, recognizing the power and authority of God. By doing so, they demonstrated their reverence for God and their commitment to upholding their word, even in challenging circumstances. God's decision to allow the Gibeonites to remain in the land, although as servants, highlights His sovereignty and mercy. Had the Israelites broken their treaty, they would have compromised their integrity and trustworthiness, aligning themselves with the deceitful practices of the people of Canaan. This serves as a powerful reminder of the importance of honesty, integrity, and honoring commitments in our interactions with others, reflecting our reverence for God and His commandments. In addition, there's a need for believers to remain steadfast in their faith and commitment to God's commandments, resisting the temptation to imitate or conform to worldly standards that contradict His ways. By staying spiritually separated and devoted to God, His people can maintain their integrity and uphold His righteousness in a world filled with spiritual impurities. This is outlined in Romans 12:2: *"Do not conform to the pattern of this world but be transformed by the renewing of your mind. Then you will be able to test and approve what God's will is—His good, pleasing, and perfect will."*

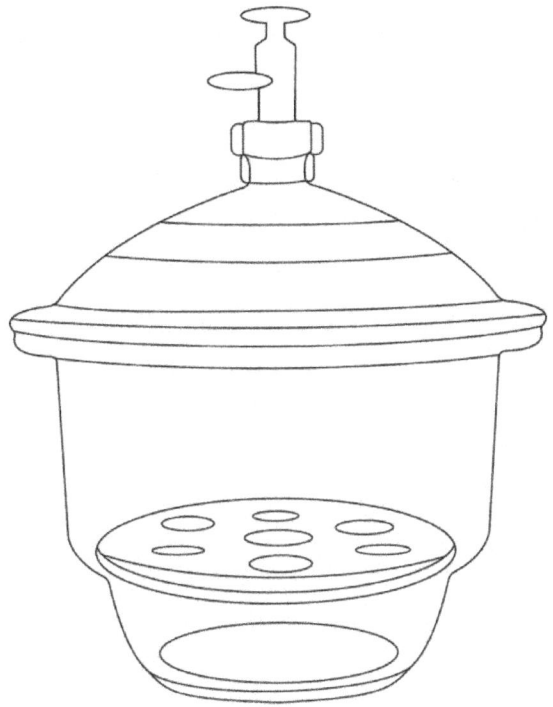

Figure 7: Desiccators are designed for the removal of moisture from materials.

Desiccator – Guardian Against Moisture

A desiccator is an essential tool in the laboratory that combats moisture. It is particularly useful when dealing with hygroscopic materials, which tend to absorb water from their surroundings. When measuring these tricky materials, the presence of water can interfere with obtaining accurate weights. But fear not! The desiccator comes to the rescue, designed to remove traces of water vapor from these moisture-loving materials, ensuring that their weight can be measured precisely without the interference of absorbed moisture.

Biblical significance: Absorbing Living Water

Just as a desiccator absorbs water vapor from hygroscopic materials, there's a beautiful analogy in the Bible, found in John 7:37-38. It says, *"If anyone is thirsty, let him come to me (Jesus) and drink. Whoever believes in Me, as the Scripture has said, streams of living water will flow from within him."* This means that when we accept Jesus into our lives and meditating on the Bible daily, we are absorbing the living water of God, which is His word. The living water purifies and nourishes our souls, much like how a desiccator removes excess moisture to maintain the purity of chemicals. By receiving this living water, we become spiritually whole and capable of pouring out God's word to others, aligning us with His divine purpose.

In the same way a desiccator restores materials compromised by moisture, the living water of God restores, strengthens us, renews our minds, and brings us peace. When we accept Jesus, we receive the Holy Spirit and have access to the other spirits of the Lord, including knowledge, wisdom, counsel, might, the fear of the Lord, and understanding (Isaiah 11:2). These gifts are treasures or resources that the Lord shares with those who wholeheartedly trust and seek to know more about Him.

In Matthew 6:33, we are encouraged to seek the knowledge of God and His kingdom: *"But seek first the kingdom of God and His righteousness, and all these things will be added to you."* When we genuinely develop a thirst to know who we are in Christ Jesus and act upon that knowledge by observing God's laws, He will supply all of our needs and grant us the desires of our hearts. This pursuit leads to noticeable physical changes in our lives, bringing relief from the challenges of the world.

Most importantly, God's living word is a powerful weapon to expose the lies of the devil, breaking and destroying his plans and control over people. Matthew 4: 2-11 illustrates the word of God used as a sword against the adversary, satan. After Jesus fasted for forty days and forty nights the devil came to tempt Him and said, *"If you are the Son of God, command that these stones become bread".* Jesus answered and said, *"It is written, 'Man shall not live by bread alone, but by every word that proceeds from the mouth of God'."* The devil temped Jesus two more times and each time Jesus consistently used the word of God as a weapon to defeat satan causing the devil to leave Him. This encounter is supported by James 4:7, which says, *"Therefore submit yourselves to God. Resist the devil and he will flee from you."*

Romans 8:17 explains that as children of God, we are heirs of God and co-heirs with Christ Jesus, giving us full access to the largest inheritance known to man. Part of our inheritance is the ability to fight the devil with the word of God in prayer. Isaiah 54:17 states, *"No weapon formed against you shall prosper; And every tongue which rises against you in judgment you shall condemn. This is the inheritance of the servants of the Lord, and their righteousness is of Me, saith the Lord."*

As sons and daughters of God, we need to pray intentionally using the word of God so that the angels of God are released to bring our prayers into fruition. Psalm 103:20 says, *"Bless the Lord, you His angels, who excel in strength, who do His word, Heeding the voice of His word."* Additionally, the word of God is a promise that will surely happen as stated in 2 Corinthians 1:20 *"For all the promises of God in Him are Yes and in Him Amen, to the glory of God through us."*

By seeking the kingdom of heaven and God's righteousness, we align ourselves with His divine plan, allowing His living water to flow into our lives. This spiritual nourishment transforms us, bringing clarity, purpose, and fulfillment. Just as the desiccator ensures the purity and effectiveness of materials in the lab, God's living water purifies our hearts and minds, making us vessels of His love and truth, ready to impact the world positively.

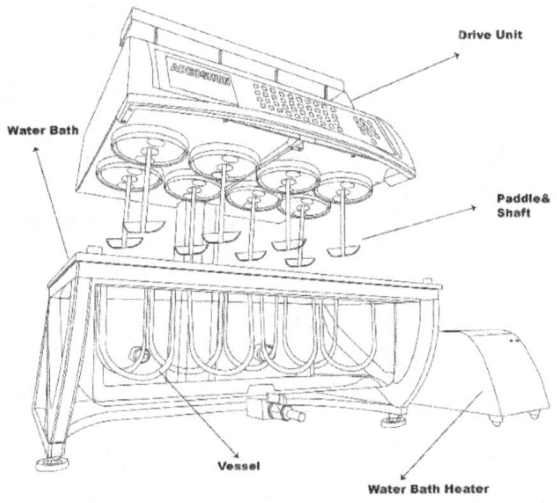

Figure 8: Dissolution testing checks how fast and how much medicine dissolves in liquid, like in your stomach.

Dissolution Testing – Ensuring Medicine's Effectiveness

Dissolution testing is like a preventative check-up for medicines. It measures how much of the active pharmaceutical ingredient (API) dissolves in a solution, ensuring that when you take a medicine, the right amount of the main ingredient that causes the desired effect dissolves into the body. This testing is crucial for tablets, capsules, ointments, gels, or patches, making sure that the medicines you use are safe and effective.

Biblical significance: Praying in Tongues – A unique connection

Similarly, praying in tongues serves as a unique and direct connection with our Father in heaven, ensuring that our prayers are specific and effective. This practice, a gift from the Holy Spirit, allows us to communicate with God in a perfect way. Even though we may not understand the words being spoken, it is a secret language between us and God, beyond the interference of the devil.

In Jude 20, believers are encouraged to pray in the Spirit, which is described as a way to build up their faith: *"But you, beloved, building yourselves up on your most holy faith, praying in the Holy Spirit."* This highlights the importance of spiritual prayer in strengthening our faith. Similarly, Psalm 141:1-2 beautifully expresses this idea: *"O Lord, I call to you; come quickly to me. Hear my voice when I call to you. May my prayer be set before you like incense; may the lifting up of my hands be like the evening sacrifice."* Praying in tongues can be seen as sending our prayers directly to God, much like the sweet fragrance of incense.

Praying in the Spirit is a powerful method for ensuring our prayers are answered because it transcends our limited understanding and faith. Matthew 7:7-8 encourages us to persist in prayer with the assurance that God is faithful to those who seek Him earnestly: *"Ask, and it will be given to you; seek, and you will find; knock, and it will be opened to you. For everyone who asks receives, and the one who seeks finds, and to the one who knocks it will be opened."*

James 4:3 warns against praying with selfish motives: *"You ask and do not receive because you ask with wrong motives, that you may spend it on your pleasures."* Praying in the Holy Spirit aligns our prayers with

God's will, preventing selfish desires from corrupting them. Romans 8:26 further explains this concept: *"In the same way, the Spirit helps us in our weakness. We do not know what we ought to pray for, but the Spirit himself intercedes for us through wordless groans."*

Faith is a vital component of effective prayer. Jesus emphasizes the importance of childlike faith in Matthew 18:2-4: *"Assuredly, I say to you, unless you are converted and become as little children, you will by no means enter the kingdom of heaven. Therefore, whoever humbles himself as this little child is the greatest in the kingdom of heaven."* Like children who trust fully and without hesitation, we should approach God with humble, unwavering faith, confident that our prayers are heard and answered. This childlike faith also includes the ability to forgive quickly and completely, maintaining a heart free of grudges and resentment.

The story of Zechariah in Luke 1:11-20, 64 serves as a powerful reminder of the importance of faith in receiving answers to our prayers. Despite receiving a direct message from the angel Gabriel, Zechariah struggled to believe, resulting in temporary muteness to prevent him from speaking against the prophecy of the child he earnestly prayed for. This illustrates that faith is essential for our prayers to be fulfilled. As stated in Matthew 21:22, *"If you believe, you will receive whatever you ask for in prayer."* When we pray in the Spirit and hold steadfast faith in God's promises, we align ourselves with His will, allowing His blessings to flow into our lives. This alignment is a profound expression of trust in God's sovereignty and His faithfulness to fulfill His word, ensuring that our prayers are answered in His perfect timing and purpose.

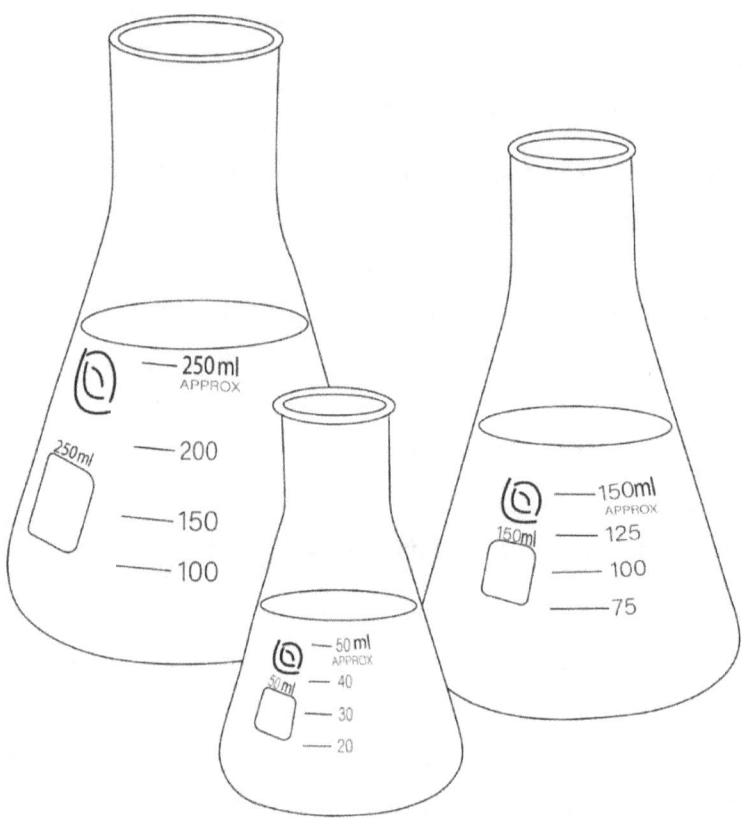

Figure 9: Erlenmeyer Flasks are used in laboratories for mixing, pouring, and storage.

Erlenmeyer Flask – A Versatile Mixing Companion

An Erlenmeyer flask is designed for easy mixing and swirling without the fear of spills. With helpful markings to measure the volume, it's a go-to tool for scientists and researchers.

Biblical significance: Metaphor of Rivers in Isaiah

In a similar way to how an Erlenmeyer flask allows easy mixing, the Lord used the metaphor of rivers to communicate a powerful message to the Prophet Isaiah. This message was a result of the people of Judah not trusting in the Lord and seeking help from Assyria against their enemies.

In Isaiah 8:7-8, the Lord describes consequences as a powerful flood passing through Judah, reaching up to the neck and swirling over the land. This imagery can be likened to the process of dissolution, where a solid material, comparable to the land of Judah, is placed into an Erlenmeyer flask, symbolizing the nation. The liquid, like a mighty river, swirls over the material, passing through it and reaching up to the neck of the flask. This represents the dissolution process, used to measure the amount of active pharmaceutical ingredient in medication, ensuring it matches dosage requirements.

Just as the Erlenmeyer flask allows controlled mixing, the metaphor in Isaiah serves as a lesson about consequences when trust is misplaced. It's a reminder to trust in the Lord and not seek solutions that might lead to unforeseen and overwhelming outcomes.

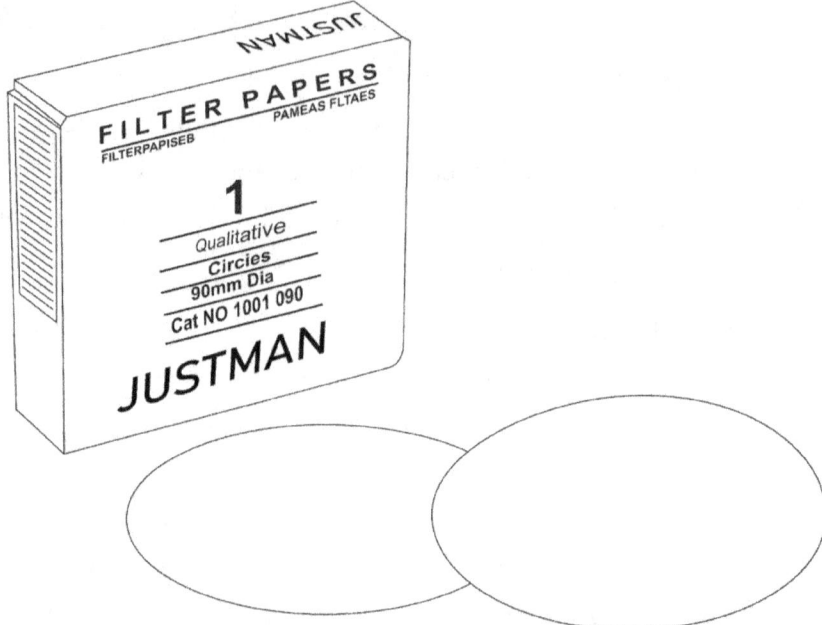

Figure 10: Filter Paper is used to separate fine solid particles from liquids.

Filter Paper and Funnels – Separating the Mix

Filter paper and funnels are essential tools for separating solids from liquids in scientific processes. The filter paper is folded and inserted into a funnel, facilitating this crucial separation. Funnels come in various sizes and materials, such as glass or plastic, with stems of varying lengths, tailored to specific tasks.

Biblical significance: Filtering Speech, Vision, Hearing and Thoughts for Positivity

Just as filter paper separates solids from liquids, Christians need filters too. We must filter what we say, see, hear, and most importantly, think. Proverbs 18:21 reminds us, *"Death and life are in the power of the tongue, and those who love it will eat its fruits."* Our words hold incredible power, shaping our reality and influencing those around us. This is why it's crucial to be mindful of our speech, ensuring we speak life and positivity.

Matthew 15:11 reveals the power of our words: *"It is not what goes into the mouth that defiles a person, but what comes out of the mouth; this defiles a person."* This scripture emphasizes the need to guard our speech, as our words can either uplift or harm. James 1:26 reinforces this idea: *"Those who consider themselves religious and yet do not keep a tight rein on their tongues deceive themselves, and their religion is worthless."* Controlling our tongue is a vital aspect of living a genuine Christian life.

As children of God, we must die to our flesh and be led by the Holy Spirit. Romans 8:5-11 explains that those who are governed by the flesh are spiritually dead because the Holy Spirit is not within them. Without the Spirit of God, they do not belong to Jesus Christ. Those led by the flesh cannot please God because they cannot submit to His law. However, when we are governed by the Spirit of God, we experience abundant life and peace through righteousness. This means living according to the Holy Spirit's guidance rather than our own fleshly desires, resulting in a life that pleases God.

This profound revelation emphasizes the necessity of aligning our words and actions with the guidance of the Holy Spirit, rather than being driven by our fleshly desires. Only by doing so can we truly belong to Christ and live in a way that pleases God, resulting in a life filled with peace and righteousness.

In John 20:29, Jesus tells Thomas, *"Because you have seen me, you have believed; blessed are those who have not seen and yet have believed."* This emphasizes the importance of filtering our perceptions and maintaining faith, even when we cannot see or fully understand the situation. It's a call to trust in God's plan and promises, even when they are not immediately visible. This is perfectly captured in 2 Corinthians 5:7: *For we walk by faith, not by sight.*

Proverbs 23:7 further reinforces this idea: *"As a man thinketh in his heart, so is he."* Our thoughts shape our character and actions. Therefore, it's crucial to filter our thinking, aiming for positivity to encourage hopeful results. Our thoughts subconsciously frame our future, physical appearance and self-esteem.

To further illustrate, consider the impact of unfiltered thoughts and words in daily life. Negative self-talk can diminish our confidence, affecting our actions and decisions. Conversely, positive affirmations can empower us, fostering resilience and success. This principle applies to our interactions with others as well. Words of encouragement can uplift and inspire, while harsh criticism can wound and demoralize.

Moreover, filtering what we see and hear is equally important. Constant exposure to negative media, gossip, or harmful content can corrupt our minds and hearts, leading us away from God's truth. Instead, we should seek content that edifies and aligns with biblical principles, filling our minds with what is noble, pure, and praiseworthy (Philippians 4:8).

By intentionally filtering our words, thoughts, and perceptions, we can cultivate a life that reflects Christ's love and righteousness.

This conscious effort not only enhances our spiritual growth but also positively impacts those around us. Just as filter paper separates impurities from a solution, we should strive for clarity and purity in our hearts and minds, fostering a life that aligns with God's will. This process is aided by inviting the Holy Spirit into our Bible reading, allowing Him to renew our minds with God's Word and reveal the hidden things we do not know, as stated in Jeremiah 33:3.

Figure 11: Flammable Cabinets are for flammable / combustible liquids to reduce risk of the contents igniting in the event of a fire.

Flammable Cabinet – Guardian Against Fire

A flammable cabinet serves as a protective barrier for flammable materials during a fire, giving emergency responders valuable time to evacuate the area and control the blaze. It is a vital safety measure that helps prevent the spread of danger.

Biblical Significance: Armor of God and Angelic Protection

In the same way, during times of spiritual attack, believers can equip themselves with the full armor of God and call upon heavenly angels for protection. Ephesians 6:11-13 encourages believers to:

"Put on the full armor of God, so that you can take your stand against the devil's schemes. For our struggle is not against flesh and blood, but against the rulers, against the authorities, against the powers of this dark world and against the spiritual forces of evil in the heavenly realms. Therefore, put on the full armor of God, so that when the day of evil comes, you may be able to stand your ground, and after you have done everything, to stand. Stand firm then, with the belt of truth buckled around your waist, with the breastplate of righteousness in place, and with your feet fitted with the readiness that comes from the gospel of peace. In addition to all this, take up the shield of faith, with which you can extinguish all the flaming arrows of the evil one. Take the helmet of salvation and the sword of the Spirit, which is the word of God."

Just as a flammable cabinet shields its contents, spiritual armor and angelic forces shield believers from the enemy's plans. Isaiah 54:17 provides further reassurance: *"No weapon formed against you shall prosper, and every tongue which rises against you in judgment you shall condemn. This is the heritage of the servants of the LORD, and their righteousness is from Me,"* says the LORD.

This means that none of the enemy's plans can succeed unless permitted by God for His greater purpose and glory. God may allow spiritual warfare to test, train, and prepare believers for their divine calling and purpose in Him. As stated in Deuteronomy 8:2, 15-16: "And you shall remember that the Lord your God led you all the way these forty years in the wilderness, to humble you and test you, to know what was in your heart, whether you would keep His commandments or not." God tests us for our ultimate good, shaping us to fulfill His will.

Moreover, calling upon heavenly angels for protection aligns with Psalm 91:11, which promises, *"For He shall give His angels charge over you, to keep you in all your ways."* These divine guardians act as our protectors, ensuring that we are shielded from harm and guided through trials.

God's allowance of spiritual warfare serves a higher purpose. It refines our faith, builds our character, and draws us closer to Him. James 1:2-4 encourages us to *"consider it pure joy... whenever you face trials of many kinds, because you know that the testing of your faith produces perseverance. Let perseverance finish its work so that you may be mature and complete, not lacking anything."* Through these trials, we gain spiritual maturity and become better equipped for the roles God has planned for us.

Just as a flammable cabinet shields materials during a fire, God promises and spiritual armor enables us to stand strong, fulfill our divine purpose, and ultimately glorify God in all that we do.

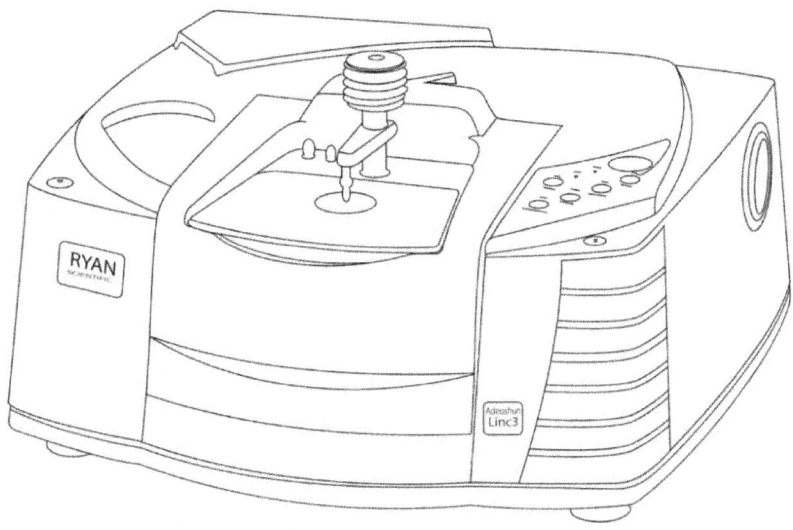

Figure 12: FTIR - An analytical technique used to characterize or identify materials.

Fourier Transform Infrared Spectroscopy (FTIR) – Illuminating Material Composition

FTIR is like a detective in the laboratory, using infrared light to unravel the mysteries of a material's composition without causing harm. This non-destructive analytical characterization technique involves passing infrared light through a sample, measuring the absorbed light at different frequencies to determine the elements present. It's like tuning into a specific radio frequency to find the right signal, revealing the unique "fingerprint" of the material.

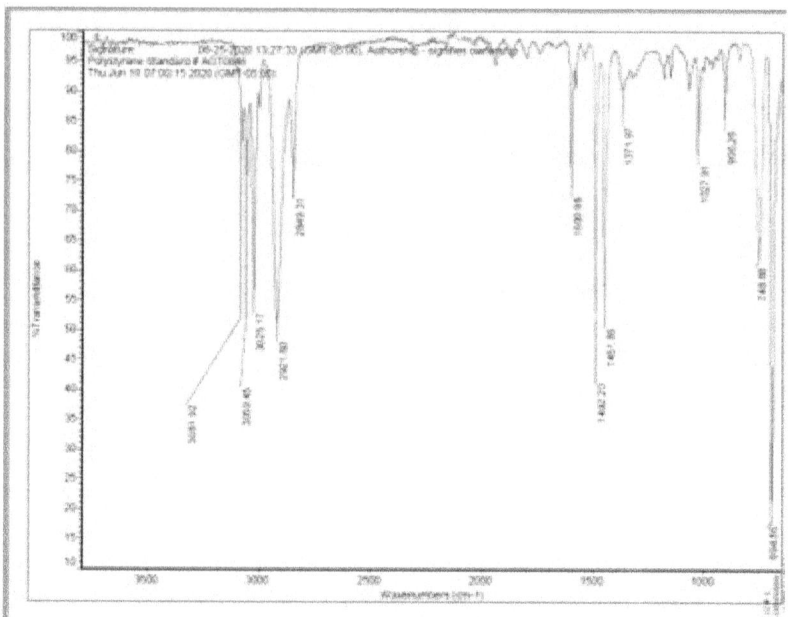

Figure 13: A spectrum of a polystyrene standard.

Biblical significance: Transforming Minds through the Word of God

Similarly, the Lord encourages us to undergo a transformative process in our lives. Just as FTIR transforms the frequencies emitted by a material into an interpretable spectrum, God wants us to shift our mindset away from worldly ways. Instead, we are urged to renew our minds by immersing ourselves in the Bible, the Word of God.

Romans 12:2 captures this idea perfectly: *"Do not conform to the pattern of this world but be transformed by the renewing of your mind. Then you will be able to test and approve what God's will is—His good, pleasing, and perfect will."* The word of God is connected to its author, the Holy Spirit of God, a living force that transforms our thinking, revealing the truth and scattering misconceptions.

Just as FTIR provides structural information about a material, engaging with the Word of God offers insight into the expectations of the Lord and reveals the plans of the enemy. The Bible is a living guide to life in real-time; it houses the laws by which we are governed and serves as one of the many ways the Lord communicates with us. The Holy Spirit, who authored the Bible and dwells within us, directs us to specific texts to reveal God's will for our current situations. When God speaks, He does so through the scriptures. Once spoken, God's words become law and are supported by the Bible. Psalm 119:72 exemplifies this, stating, *"The law of Your mouth is better to me than thousands of gold and silver pieces."* This alignment allows us to synchronize our actions with His will, creating a spiritual spectrum that guides us toward a good, pleasing, and fulfilling life.

The transformative power of God's Word renews our minds, allowing us to see the world from a divine perspective. This renewal process helps us identify and reject worldly patterns that lead us astray. Instead, we embrace God's truth, which shapes our character and directs our actions. This ongoing transformation is essential for spiritual growth and aligning ourselves with God's purposes.

Moreover, the Bible provides a comprehensive understanding of God's expectations and the enemy's strategies. By regularly studying and meditating on scripture, we equip ourselves with the knowledge and wisdom needed to navigate life's challenges. The

Holy Spirit uses the Word of God to guide us, offering comfort, conviction, and direction. This divine guidance ensures that we stay on the path that leads to a fulfilling and abundant life in Christ.

In conclusion, just as FTIR transforms material frequencies into valuable information, the Word of God transforms our minds, revealing divine truth and guiding our lives. By immersing ourselves in scripture and allowing the Holy Spirit to lead us, we align with God's will and experience the abundant life He promises. This transformation not only benefits us personally but also enables us to impact others positively, reflecting God's love and truth in a world in need.

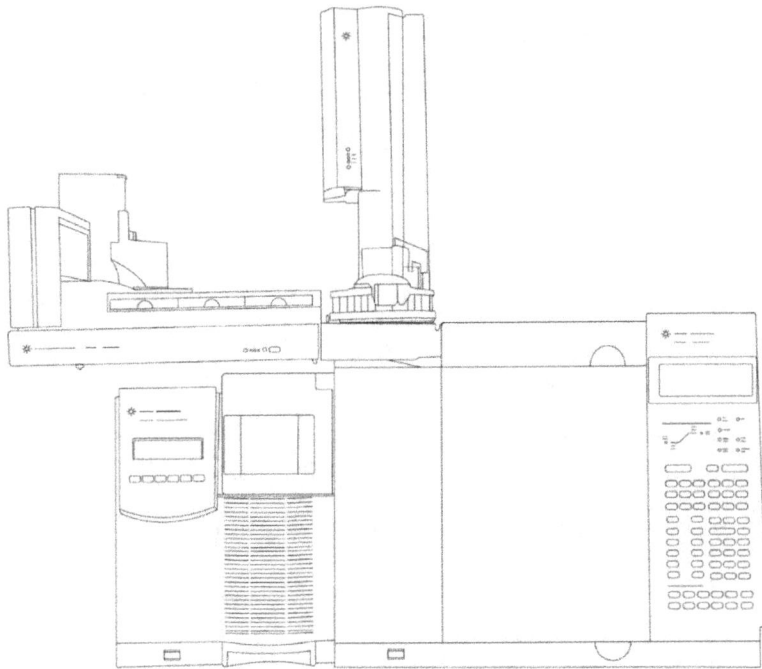

Figure 14: GC-MS is a combination of two analytical techniques to identify substances in a sample.

Gas Chromatography Mass Spectrometry (GC-MS) – Unraveling Molecular Mysteries

Gas Chromatography Mass Spectrometry (GC-MS) is a combined instrument used to separate and identify compounds within a mixture. It utilizes a gas gradient, a pressure change over time, to push gaseous analytes through a column and into the detector for identification.

To begin, a uniformly mixed solution with a known concentration is added to a GC vial. The vial is then loaded onto the autosampler

47

of the instrument, and a sample portion is injected into the instrument's inlet. The inlet, heated to a specific temperature, vaporizes the compounds or analytes, turning them into gas. The gaseous analytes are then propelled into the instrument's column using pressure changes created by a carrier gas such as helium, nitrogen, or hydrogen. Once the gases enter the column, the separation process begins.

Inside the column, a chemical coating forms the stationary phase. Analytes interact with this coating, with some spending more time than others, similar to social interactions. Those spending less time, called early eluting peaks, reach the detector faster, while those spending more time, termed late eluting peaks, take longer. These peaks are identified by their retention time, indicating the amount of time interacting with the column post-injection. This data is depicted in a chromatogram, showing peak retention time and intensity counts, reflecting the amount of compound reaching the detector.

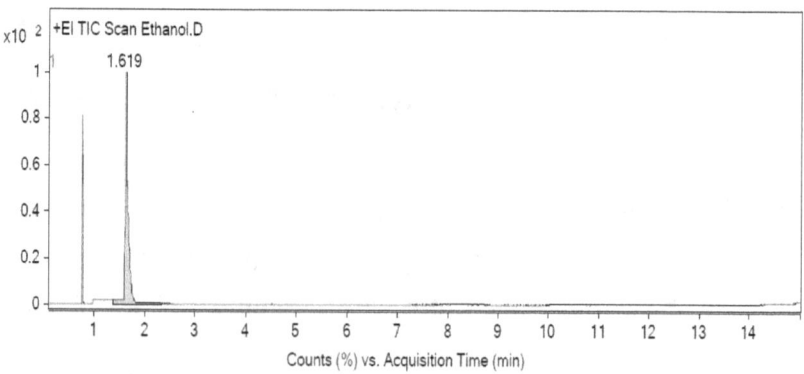

Figure 15: Total Ion Chromatogram (TIC) of a sample.

In the second phase, analytes are transferred through the transfer line into the Mass Spectrometry (MS) section of the instrument. Once inside the MS, analytes undergo ionization, where neutral molecules from the sample are converted into charged ions by losing electrons. This ionization gives the molecules a positive charge, allowing for separation based on their size or mass in the quadrupole. The quadrupole guides the positively charged ions to the Mass Spectrometry Detector (MSD), where they are identified.

The chromatogram displays the fragmentation pattern, mass-to-charge ratio (m/z), and intensity counts of the compound. The intensity counts represent the amount of analyte detected, with the signal being amplified based on the analyte's concentration. The mass-to-charge ratio (m/z) reflects the relationship between an ion's mass and its charge. The fragmentation pattern illustrates how the analyte breaks down in the detector. This data is compared to a library of known chemical fragmentation patterns to identify compounds. Alternatively, the operator can analyze a mixture of pure reference standards, using retention time, mass-to-charge ratio, and fragmentation pattern to match and identify compounds in the sample.

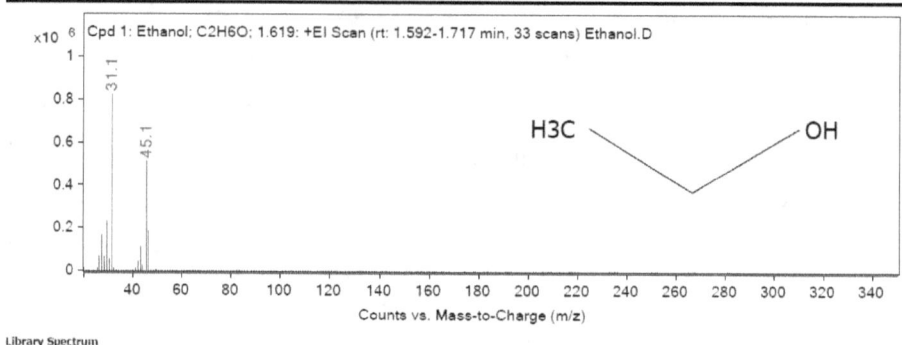

Library Spectrum

Figure 16: The fragmentation pattern of ethanol.

Biblical significance: Faith and Trust in God's Provision

While equipment like this may seem complex, much like life, faith in Jesus can simplify things. When we trust in Him and believe in what we're asking for, aligning with God's will, He provides. Philippians 4:19 reassures believers, *"But my God shall supply all your need according to his riches in glory by Christ Jesus."* This emphasizes the belief that God can provide for our needs according to His divine plan.

Hebrews 11:1 defines faith as having confidence in the unseen, and even a mustard seed-sized faith, as mentioned in Matthew 17:20, can achieve the seemingly impossible. This illustrates the power of belief and trust in God's ability to bring about what we hope for.

James 2:26 reinforces the idea that faith must be accompanied by action. It states, *"For as the body without the spirit is dead, so faith without works is dead also."* This highlights the importance of backing up our faith with actions, taking tangible steps towards the manifestation of our hopes and aspirations.

Fundamentally, just as GC-MS unravels molecular mysteries, our faith, actions, and trust in God unravel life's complexities, offering a guiding light on our journey.

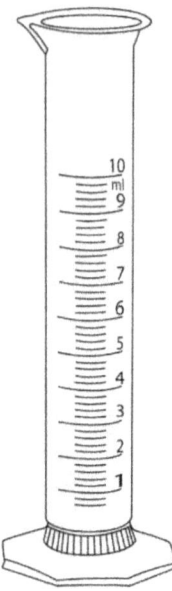

Figure 17: A Graduated Cylinder is used to measure the volume of a liquid.

Graduated Cylinder – Precision in Measurement

Graduated cylinders are essential tools in laboratories, primarily used to measure the volume of liquids accurately. They feature markings to indicate volume and spouts on their rims to aid pouring. Available in various sizes, smaller diameters offer more precise volume measurements. When reading the volume from a graduated cylinder, the liquid's surface appears to have a curved shape, called the meniscus. To accurately measure, align the lowest point of the meniscus with the nearest marking, ensuring the cylinder is level. This technique ensures precise volume readings.

Biblical significance: Honesty in Measurement

The importance of accurate measurement is echoed in passages from the Bible, such as Deuteronomy 25:15-16. It states, *"You must have accurate and honest weights and measures, so that you may live long in the land the Lord your God is giving you. For the Lord your God detests anyone who does these things, anyone who deals dishonestly."*

This Biblical significance aligns with the principles of using tools like graduated cylinders, and balances. When used correctly and honestly, these instruments reflect accuracy and precision, which is in accordance with the biblical teachings emphasizing truthfulness in transactions. In essence, the graduated cylinder, when employed with integrity, becomes a symbol of the importance of honesty in measurement and is deemed acceptable in the sight of God. This reinforces the idea that conducting activities with accuracy and truthfulness aligns with moral values upheld in the Kingdom of God.

Figure 18: Hot Plates heat materials and perform chemical reactions.

Hot Plate – Flameless Catalyst for Reactions

Hot plates are like special stoves used in labs. Instead of fire, they make things hot without flames. Scientists use them to mix and heat up chemicals safely. Hot plates help keep experiments under control.

Biblical significance: Controlling Emotional Reactions

Similar to how hot plates control chemical reactions, individuals have the capacity to control their emotional reactions, especially in moments of disappointment or anger. When faced with frustration, the body can undergo a "chemical reaction" of stress, releasing hormones like adrenaline and cortisol, encouraging a reaction.

Consequently, individuals may express physical or emotional distress toward the source of their anger.

Indeed, God calls us to resist the temptation of acting on our immediate impulses and instead seek His guidance for our next steps. By relinquishing control to Him, we allow God to resolve situations in the best possible way. This demonstrates trust in His wisdom and ability to navigate challenges with grace and purpose. Ephesians 4:31-32 advises against letting anger and bitterness take control: *"Get rid of all bitterness, rage, and anger, brawling and slander, along with every form of malice. Be kind and compassionate to one another, forgiving each other, just as in Christ, God forgave you."* This encourages individuals to control emotional "flames" and react with kindness, compassion, and forgiveness, aligning with God's teachings.

God sets the standard for how we should behave and react. This comparison between the controlled environment of a hot plate and the controlled emotional response advocated in Ephesians emphasizes the importance of aligning our actions with the higher standard of God's teachings.

Figure 19: Incubators are used to control humidity, temperature, and other conditions to grow and preserve cell and microbiological cultures.

Incubator – Nurturing Growth in the Word

An incubator, in the scientific realm, creates a temperature-controlled environment to nurture the growth of microorganisms. Similarly, a moral analogy can be drawn, suggesting that immersing oneself in the word of God provides the optimal environment for personal and spiritual growth.

Biblical significance: Incubating in the Word of God

The secret to a successful life, as metaphorically expressed, is to incubate in the word of God. Just as an incubator offers the right temperature for microorganisms to flourish, immersing ourselves in the teachings of God creates an environment conducive to personal and spiritual development.

Galatians 5:22-23, as shared by Apostle Paul, outlines nine characteristics that manifest in the lives of individuals filled with the Spirit of God: love, joy, peace, forbearance, kindness, goodness, faithfulness, gentleness, and self-control. This emphasizes the transformative power of God's word, acting as a spiritual incubator that nurtures positive attributes and virtues within individuals.

In principle, just as an incubator supports the growth of microorganisms, immersing oneself in the word of God provides the optimal conditions for the growth of morals and qualities that contribute to a fruitful and fulfilling life.

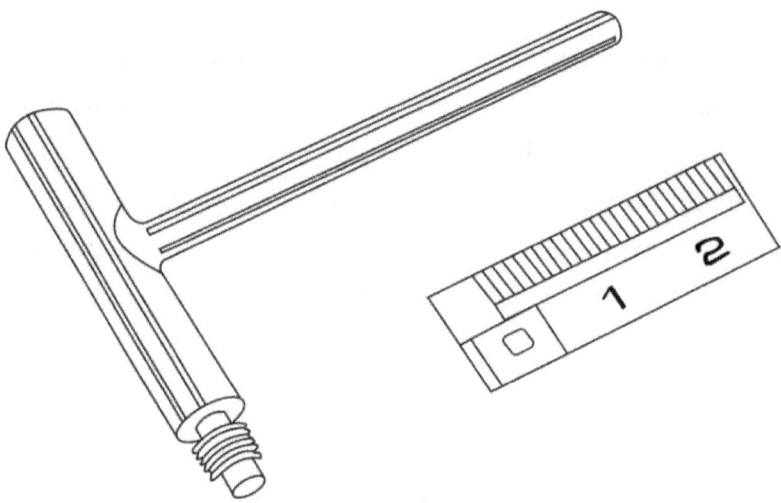

Figure 20: Jig - A tool used to measure and set the correct column length into a GC-MS system.

Jig – Precision in Instrumentation

A jig serves as a measuring tool in laboratories, specifically designed for setting the correct column length into the Gas Chromatography (GC) Injector, ensuring accurate measurements for optimal instrument performance.

Biblical significance: The Significance of Accurate Measurements

Accurate measurements are vital in various applications, including the careful setup of equipment. In the context of a GC-MS system, correctly installing a column into the detector is crucial to avoid issues like chemical over-saturation, which can lead to potential instrument downtime for maintenance.

The word of God serves as a measuring rod for us to gauge and reflect the values of God in our lives and individual purpose on earth. Revelation 11:1 provides a biblical perspective on the significance of measurements. In this verse, John is given a reed like a measuring rod and instructed, *"Go and measure the temple of God and the altar, with its worshipers."* Measuring the temple is attributed to both judgment and preservation by God, whereby the righteous and unrighteous will be judged but only the righteous will be preserved.

In principle, just as a jig ensures the correct installation of laboratory equipment, the biblical reference emphasizes the significance of measurement in spiritual contexts by God. This highlights the value of precision and accuracy, both in scientific endeavors and in matters of faith.

Figure 21: Karl Fischer Titrator measures the amount of water in a sample.

Karl Fischer Titrator – Assessing Life's Essential Element

A Karl Fischer titrator is a laboratory instrument used to precisely measure the amount of water in various products. This method ensures accurate monitoring of moisture levels, which is crucial for maintaining the quality and stability of pharmaceutical products, as even small amounts of water can significantly affect their efficacy and safety. This process is especially important considering that water, a fundamental substance essential to life, is present in all living organisms.

Biblical significance: Water as a Symbol of Life and the Holy Trinity

Water serves as a symbol of life, representing purity and necessity for existence. This concept is mirrored in the Holy Trinity, drawing parallels between the three phases of water and the Father, Son, and Holy Spirit.

In Psalm 62:6, David characterizes God as a rock, stating, *"He alone is my rock and my salvation; he is my fortress, I will not be shaken."* This aligns with the solid phase of water, symbolizing the stability and strength found in God.

In John 4:10, Jesus is represented as the living waters: *"Jesus answered her, 'If you knew the gift of God and who it is that asks you for a drink, you would have asked him, and he would have given you living water.'"* This parallels the liquid phase of water, signifying the life-giving and purifying nature of Jesus.

The gift of the Holy Spirit, analogous to water vapor, is documented in Acts 2:38: *"And Peter said to them, 'Repent and be baptized every one of you in the name of Jesus Christ for the forgiveness of your sins, and you will receive the gift of the Holy Spirit.'"*

The Holy Trinity is summarized in Matthew 28:19: *"Therefore go and make disciples of all nations, baptizing them in the name of the Father and of the Son and of the Holy Spirit."* This reflects the unity and interconnectedness of the Father, Son, and Holy Spirit, akin to the inseparable phases of water.

Fundamentally, the Karl Fischer titrator, assessing water content, draws a moral equivalence to the symbolism of water in the Holy Trinity, emphasizing the essential and symbolic nature of water in both the physical and spiritual realms.

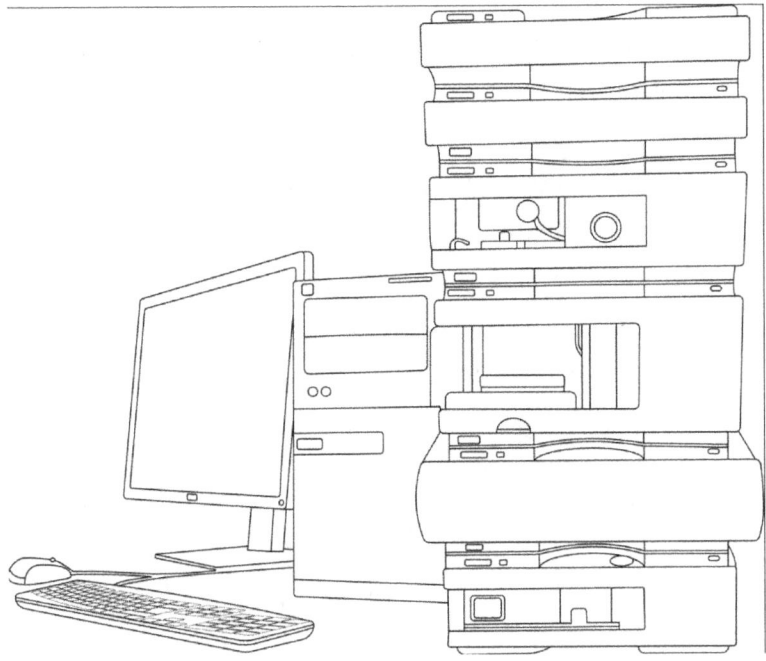

Figure 22: LC-MS is used to separate thermally labile compounds in a chemical mixture.

Liquid Chromatography Mass Spectrometry (LC-MS) – Spiritual and Analytical Separation

Just like a GC-MS, Liquid Chromatography Mass Spectrometry (LC-MS) is a characterization technique serves as a dual-purpose instrument, separating mixtures to identify compounds. In a similar manner, God's grace serves as a spiritual separator, shielding believers from the consequences of sin and guiding them towards the abundant life promised by Jesus.

Biblical significance: Spiritual Separation and Protection

LC-MS employs a solvent gradient for separation, mirroring God's grace, which acts as a spiritual gradient separating believers from the consequences of sin. The injection of a mixed solution into the instrument resonates with believers navigating a world filled with challenges and temptations. As analytes interact with the stationary phase of the column, drawing parallels to believers interacting with the world, early eluting peaks symbolize those with less connection to worldly influences, while late eluting peaks represent individuals spending more time in the world.

The second phase involves analytes reaching the source detector, desolvating into the gaseous phase, and fragmenting into charged particles. Similarly, believers undergo spiritual refinement through encounters with the Holy Spirit. The chromatogram, displaying retention time and intensity counts, mirrors the unique spiritual journey with God.

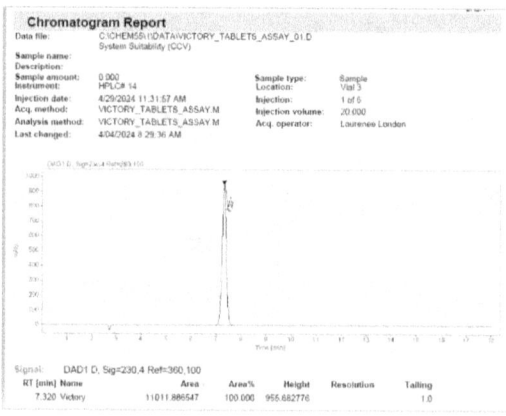

Figure 23: TIC of an active pharmaceutical ingredient (API) standard material.

John 10:10 serves as a biblical anchor, highlighting satan's desire to steal, kill, and destroy, countered by Jesus' promise of abundant life for believers. Satan targets God's children due to their unique status, made in God's image, given dominion over all of God's creations, and capable of repentance. Satan's legal right to attack is obtained through sin, whether personal, territorial, or foundational. Personal sin, as outlined in 1 John 1:8, requires repentance and forgiveness. Territorial sin, exemplified by Sodom and Gomorrah or Nineveh, showcases God's judgment on regions entrenched in wickedness. Foundational or bloodline sin, explained in Psalm 11:3 and Exodus 20:4-6, reveals generational consequences. Yet, God's redemptive power allows believers to repent for their sins, their ancestors' sins, and intercede for territories. Leviticus 26:40-42 emphasizes God's willingness to forgive us of our sins and heal our land upon genuine repentance.

At its core, LC-MS serves as a metaphor for God's spiritual separation and protection, symbolizing the intricate journey of believers with their unique experiences. The biblical significance is found in the scriptural promises of abundant life, redemption, and healing through repentance and faith in God.

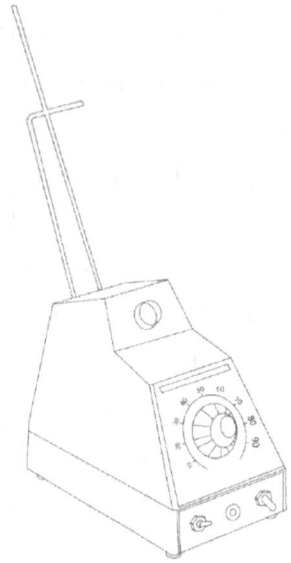

Figure 24: A Melting Point Apparatus is used to determine the melting point of a substance.

Melting Point – Divine Judgment and Righteousness

The melting point of a substance is the temperature at which it transitions from a solid to a liquid. In a spiritual analogy, Psalm 68:1-2 portrays God reaching His metaphorical "melting point," where His divine anger causes His enemies to scatter. The passage reads, *"Let God arise, let His enemies be scattered; let those also who hate Him flee before Him. As smoke is driven away, so drive them away; as wax melts before the fire, so let the wicked perish at the presence of God."* This scripture carries a moral lesson, highlighting the importance of righteous living to avoid becoming enemies of God.

Biblical significance: Divine Judgment and Righteous Living

Aligning one's life with God's principles and seeking His favor ensures protection from divine judgment. The analogy of melting wax emphasizes the vulnerability of those who oppose God, urging individuals to live in a manner that pleases Him. Living a life pleasing to the Lord ensures that one does not become satan's prey or the recipient of God's righteous anger.

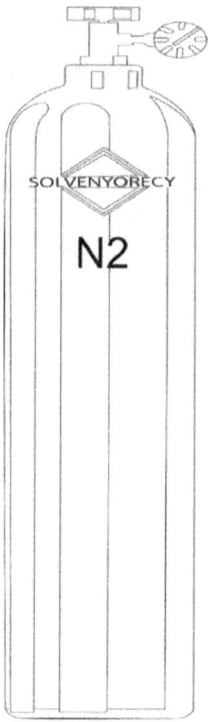

Figure 25: Nitrogen Tank stores nitrogen an inert gas commonly used to create a controlled, oxygen free environment.

Nitrogen Tank – Empowering Presence of the Holy Spirit

A nitrogen tank, containing the inert gas nitrogen, serves various scientific purposes, often acting as a carrier gas in analytical instruments like GC-MS and LC-MS. Nitrogen's inert nature, its resistance to chemical reactions with other substances, signifies its purity, enabling it to remain in the gas phase.

The purity of nitrogen gas draws a parallel to the empowering and unchanging presence of the Holy Spirit in the life of a believer. Just as nitrogen maintains its integrity and purity, the Holy Spirit empowers and remains steadfast in guiding and supporting believers of Jesus Christ.

Biblical significance: The Holy Spirit's Empowering Presence

In Luke 24:49, Jesus, before ascending to heaven, promises to send the Holy Spirit as a guiding and empowering presence for believers. This spiritual similarity draws on the characteristics of nitrogen—non-reactive and maintaining its purity. The Holy Spirit, like nitrogen, is unwavering in purity, providing believers with power, understanding, and gifts for effective service as Ambassadors of Christ Jesus on earth. As Ambassadors of the Lord, like disciples, it is our duty to spread knowledge about the Kingdom of God, as stated in Luke 9:1-2: *"And he called the twelve together and gave them power and authority over all demons and to cure diseases, and he sent them out to proclaim the kingdom of God and to heal."*

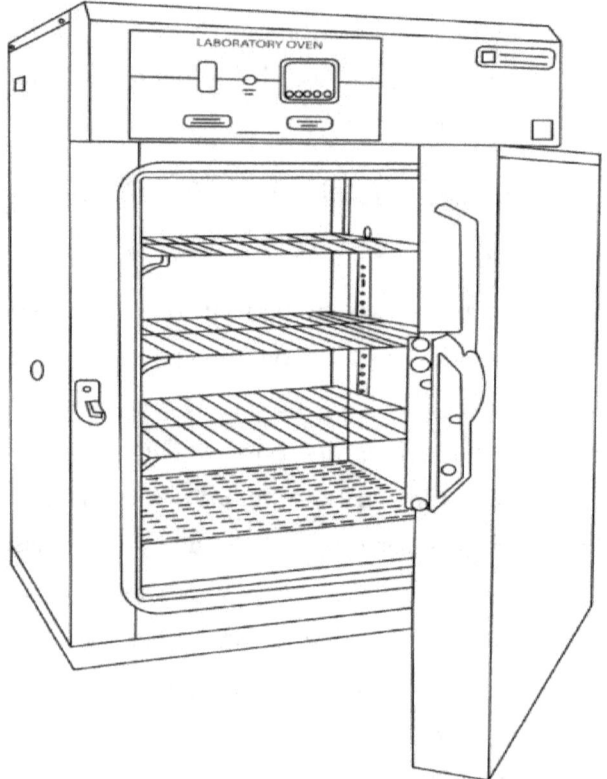

Figure 26: Ovens are used for drying, sterilization, heating, evaporation and more in science laboratories.

Oven – A Source of Drying

Laboratory ovens, often utilized for drying materials or glassware, find an intriguing similarity in Isaiah 44:15, where an oven becomes a metaphor for the mindset of an idolater. This scriptural analogy serves as a warning against the worship of inferior gods or idols.

Biblical significance: Idolatry and False Worship

The Bible presents a compelling contrast between the true God, Elohim, and false gods through the metaphor of wood used for different purposes. In one biblical passage, wood serves practical and beneficial functions: people use it to bake bread and to keep warm, both essential and good uses. However, from the same wood, some fashion idols—statues meant for worship—which God condemns. While the wood used for sustenance provides real help, the idols made from that same wood are powerless, deceptive, and ultimately harmful.

God emphasizes that He alone is the true and living God, deserving of worship and devotion. Unlike false gods or idols, which offer the illusion of help but conceal their destructive nature, Elohim is the only one who genuinely cares for, protects, and sustains His people. The Bible warns against the worship of idols, because, although they might appear to offer assistance or blessings, they cannot deliver the way the Almighty God can. Instead, they deceive, drawing people away from truth and life.

When a person recognizes the emptiness of idol worship and turns toward the living God, the false god's true destructive nature is often revealed. No longer able to hide behind a façade of usefulness, the false god may retaliate, revealing its malevolence in the face of rejection. This stark contrast highlights the reality that Elohim is the complete antithesis of false gods: He is alive, powerful, and compassionate. He offers genuine relationship and protection, and unlike false gods, He does not force His worship on anyone. He respects human free will, giving each person the choice to serve Him or not.

Joshua 24:15 encapsulates this choice: *"And if it seems evil to you to serve the Lord, choose for yourselves this day whom you will serve, whether the gods which your fathers served that were on the other side of the River, or the gods of the Amorites, in whose land you dwell. But as for me and my house, we will serve the Lord."* Joshua's declaration emphasizes personal choice and commitment to serve the true God, acknowledging the presence of alternatives but affirming the superiority and truth of serving Elohim.

This passage reflects the critical decision each person faces: whether to pursue false, destructive idols or to follow the one true God, who offers life, peace, and genuine care.

Isaiah 44:6 boldly declares, *"This is what the LORD says—Israel's King and Redeemer, the LORD of hosts: I am the first and I am the last; beside me there is no God."* This profound statement emphasizes the supremacy and uniqueness of God, affirming that no other deity or power can rival Him. This truth is vividly illustrated in the narrative found in Numbers 22:1-21, where Balak, the king of Moab, becomes fearful of the approaching Israelites and seeks to hire Balaam, a sorcerer, to curse them.

Balak's intent to use sorcery against God's chosen people demonstrates the desperation and fear he felt as Israel, under God's protection, moved closer to his kingdom. However, despite Balak's efforts, Balaam is instructed by God not to curse the

Israelites. God's direct intervention reveals a powerful truth: even those who oppose God's people must submit to His will. Balaam, though a pagan diviner, recognizes God's authority and ultimately obeys, refusing to curse the Israelites as Balak had requested.

This narrative is significant because it underscores the reality that no adversary can harm God's people without His permission. The story of Balaam and Balak highlights God's sovereign control over all circumstances, even those involving dark forces or hostile intentions. In Numbers 23:8, Balaam himself testifies to this truth when he says, *"How can I curse those whom God has not cursed? How can I denounce those whom the LORD has not denounced?"* This declaration reflects the divine protection that surrounds God's people.

When believers live in repentance, holiness, and righteousness, they align themselves with God's will and receive His protection. The enemy may attempt to attack or bring harm, but just as Balaam was unable to curse Israel, these efforts will consistently fail. God's favor and blessing rest upon those who walk uprightly before Him, making them untouchable by the schemes of the enemy.

This story highlights the vast difference between the false gods Balak relied on and the one true God, who cannot be manipulated or challenged. God's power is absolute, and His commitment to His people is unwavering. As Isaiah 44:6 declares, there is no other God besides Him—He alone is the beginning and the end, the first and the last. This truth offers profound reassurance to all who trust in Him, knowing that they are under the care and protection of the Almighty, and no curse or force of darkness can prevail against them.

In summary, the lesson emphasizes the importance of trusting and worshipping the one true God, the creator of all things, who looks after us and is greater than any other god including our enemies. It advises against worshipping false gods or prioritizing other things over Him. The moral takeaway urges believers to stay committed to worshipping the living God, recognizing His authority and power.

Figure 27: PPE is gear used in science labs and other work environments to protect workers from hazardous chemicals and other potential dangers.

Personal Protective Equipment (PPE) – Protection in the Physical and Spiritual Realm

Personal Protective Equipment (PPE) refers to gear worn to reduce exposure to potential hazards that could lead to severe injuries, infections, or illnesses. This equipment typically includes a lab coat, safety goggles, gloves, closed-toe shoes, and head protection. Interestingly, PPE draws a compelling parallel to the protection believers require in the spiritual realm, often referred to as the "armor of God."

Biblical significance: Armor of God as Spiritual Protection

Ephesians 6:11-13 (NIV): "Put on the full armor of God so that you can take your stand against the devil's schemes. For our struggle is not against flesh and blood, but against the rulers, against the authorities, against the powers of this dark world and against the spiritual forces of evil in the heavenly realms. Therefore, put on the full armor of God so that when the day of evil comes, you may be able to stand your ground, and after you have done everything, to stand. Stand firm then, with the belt of truth buckled around your waist, with the breastplate of righteousness in place, and with your feet fitted with the readiness that comes from the gospel of peace. In addition to all this, take up the shield of faith, with which you can extinguish all the flaming arrows of the evil one. Take the helmet of salvation and the sword of the Spirit, which is the word of God."

This scriptural reference reinforces the importance of spiritual readiness through the armor of God. The full armor of God includes various components such as the belt of truth, the breastplate of righteousness, the shoes of the gospel of peace, the shield of faith, the helmet of salvation, and the sword of the Spirit. The spiritual armor serves as a means of defense against the unseen spiritual forces that seek to harm believers. As believers our struggle is not solely physical but deeply spiritual. Just as physical PPE protects individuals from harm, the armor of God serves as the ultimate spiritual protection against the schemes of the devil.

In summary, by putting on the full armor of God daily, believers are equipped to stand firm in their faith and withstand the attacks of the enemy. This armor empowers believers to navigate the life's journey with resilience and confidence, knowing they are protected by God's strength and guidance.

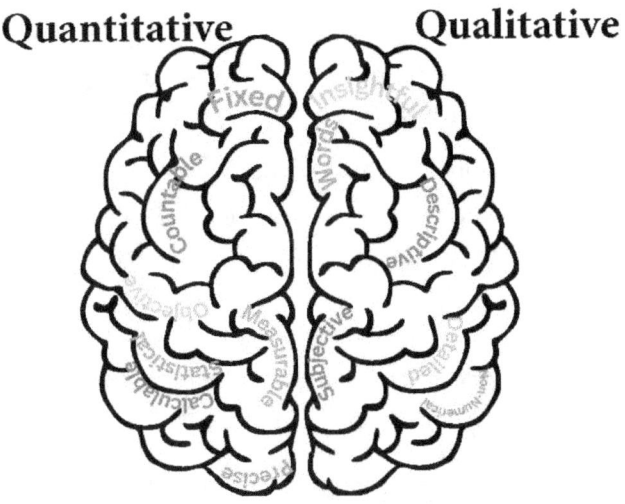

Figure 28: Qualitative data describes characteristics and qualities using non-numerical categories, while quantitative data involves numerical measurements and counts that can be statistically analyzed.

Qualitative/Quantitative Data: Faith Measurement through Jesus' Teachings

Qualitative data is often descriptive and conceptual, categorizing properties and attributes with words instead of numbers. In a biblical context, Jesus frequently used qualitative comparisons to convey the measure of faith. His teachings highlight the varying degrees of faith, portraying it as something that can be great or small. For instance, in Matthew 15:28, Jesus acknowledges a woman's "great faith," emphasizing the qualitative aspect. Conversely, in Matthew 8:26, he addresses the disciples' "little faith." These qualitative measurements emphasize the dynamic nature of faith and its varying degrees of strength.

Quantitative data involves numerical measurements, and the Bible provides instances where specific measurements and units are employed. In Matthew 26:14-15, the betrayal of Jesus by Judas Iscariot involves a quantitative transaction of thirty silver coins. This specific count illustrates the quantifiable nature of the exchange. Additionally, when God instructs Noah to build the ark in Genesis 6:15, the dimensions (450 feet long, 75 feet wide, and 45 feet high) illustrate quantitative specifications for the construction.

Biblical significance: Faith as a Developmental and Measurable Aspect

The Bible showcases the measurable aspects of faith. Faith, as depicted in the Bible, can grow or diminish. Romans 10:17 emphasizes the role of hearing the word of God in the development of faith. Living by faith (2 Corinthians 5:7) and the impossibility of pleasing God without faith (Hebrews 11:6) underline faith as a vital and measurable element in the Christian walk.

In summary, the utilization of both qualitative and quantitative data aids in conveying biblical teachings, enriching the comprehension of concepts such as faith and practical instructions for building or payment. Scriptures employ a blend of descriptive and numerical language to communicate measurements and dimensions relevant to the biblical context. This multi-layered approach enhances the depth of understanding and simplifies the application of biblical principles in daily life.

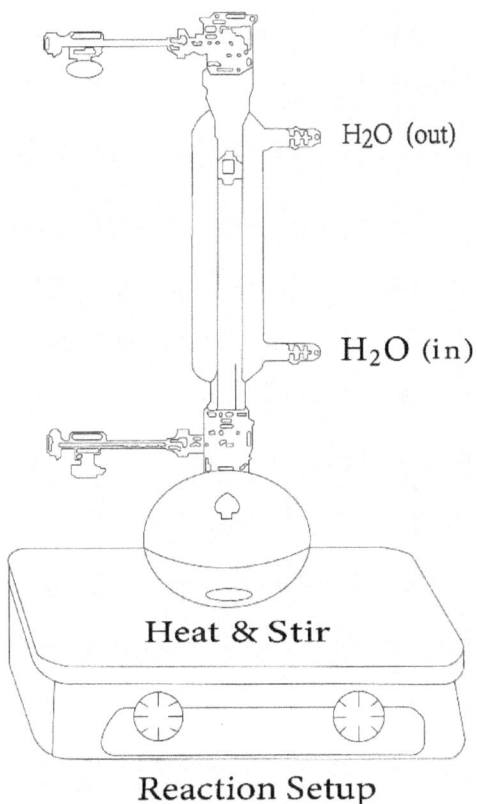

H₂O (out)

H₂O (in)

Heat & Stir

Reaction Setup

Figure 29: The recrystallization method is a technique used to purify solid compounds by dissolving them in a hot solvent and then allowing them to slowly crystallize as the solution cools.

Recrystallization: The Purification Process

Recrystallization is a purification method used for chemicals based on their solubility in a suitable solvent. Solubility refers to a substance's ability to dissolve in a solvent, such as water. The process involves heating the material to dissolve the compound with impurities, followed by filtering the solution to separate the purified product from other materials.

Biblical significance: Purification Through Repentance

The recrystallization technique, with its purification process, draws a connection to the spiritual concept of repentance and purification through the sacrifice of Jesus Christ.

When individuals commit sins, they separate themselves from God and become adversaries, as sin is incompatible with God's design for a righteous life. Romans 6:23 highlights the consequence of sin, stating, *"The wages of sin is death, but the gift of God is eternal life in Christ Jesus our Lord."* Sin affects various aspects of life, leading to spiritual, physical, mental, and emotional consequences, causing internal conflict and even health issues.

The Biblical significance lies in God's love and the provision of a solution for purification. God, out of love, sent His Son as a living sacrifice to atone for our sins. Through the blood of Jesus, as Acts 2:38 emphasizes, repentance and forgiveness are made possible. The recrystallization technique, used for the purification of chemicals, becomes a metaphor for the purification of sins through the blood of Jesus.

The cleansing power of the blood of Jesus is emphasized in John 1:7, stating, "The blood of Jesus His Son cleanses us from every sin." Just as recrystallization isolates the purified product from impurities, the blood of Jesus purifies individuals from the stain of sin, restoring the connection and reconciliation with God.

In essence, the recrystallization technique becomes a physical reminder of the spiritual process of repentance and purification through the redemptive power of the blood of Jesus, highlighting God's love, forgiveness, and the path to restoration.

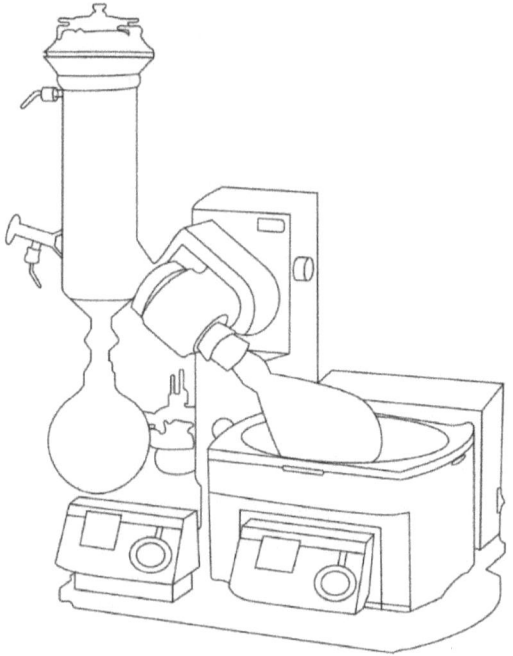

Figure 30: A rotary evaporator is a lab tool that removes solvents from a solution by heating and spinning the liquid, making evaporation faster and more efficient.

Rotary Evaporator Process and Spiritual Analogy

A Rotary Evaporator is an essential piece of equipment commonly used in chemical laboratories to efficiently remove solvents from a mixture through evaporation. The process starts by placing the liquid sample into a round-bottom flask, which is then attached to the evaporator system. The flask is partially submerged into a heated water bath, and as the system operates, the flask rotates continuously. This rotation, combined with the application of a vacuum, lowers the atmospheric pressure within the system. As a result, the liquid evaporates at a much lower temperature than its usual boiling point, making the process more efficient and reducing the risk of heat-sensitive substances being damaged.

The evaporated liquid turns into vapor, which is directed into a condenser where the vapor is cooled down and transformed back into a liquid—a process known as condensation. The condensed liquid, which is typically the solvent, is collected in a separate flask, allowing the desired product or material to remain in the original flask, separated from the solvent it was synthesized in.

This process is highly effective in isolating compounds from solvents, especially when dealing with sensitive or volatile substances. The rotary evaporator's ability to evaporate liquids at lower temperatures makes it ideal for applications such as solvent recovery, concentration of solutions, and purification of reaction products in chemical research and production.

Biblical significance: Rotary Evaporator Process and Prayer in the Spirit Analogy:

1. **Evaporation by Creating a Vacuum:**

    o *Rotary Evaporator:* The vacuum created in the rotary evaporator facilitates evaporation at a lower temperature than usual.

    o *Spiritual Analogy:* When believers pray from the third heaven in the Spirit, it symbolizes creating a spiritual vacuum. Praying in tongues (praying in the Spirit) is often seen as a direct communication with God, allowing for a unique and profound connection.

2. **Condensation and Collection:**

   o *Rotary Evaporator:* The evaporated liquid undergoes condensation, turning back into a liquid, which is then collected in another flask.

   o *Spiritual Analogy:* Similarly, in the spiritual realm, there is a process of condensation when prayers ascend and interact with God. Psalm 141:2 illustrates this concept by likening our prayers to incense set before the Lord and the lifting up of our hands to the evening sacrifice. The response, guidance, or blessings from God are metaphorically collected and manifested in the lives of believers.

3. **Entering the Spiritual Vacuum:**

   o *Rotary Evaporator:* The rotary evaporator creates a controlled environment for evaporation and condensation.

   o *Spiritual Analogy:* Believers are encouraged to pray in the Spirit, creating a spiritual atmosphere or vacuum by allowing the Holy Spirit to intercede for them through powerful wordless groans. This enables believers to pray with divine intention, targeting specific areas of need. Direct communication with God in this manner aids in receiving answered prayers because satan cannot comprehend our individual prayer language, therefore he cannot hinder or block our prayers when we pray in the Spirit.

In summary, the operation of a rotary evaporator, with its process of evaporation and condensation, parallels the spiritual journey believers undergo to connect with the throne room of God. When believers pray in the Spirit from the third heaven, they engage in direct communication with Jesus, creating a sacred space where natural and spiritual barriers are transcended. As Ephesians 2:6 states, believers are raised up and seated together in heavenly places in Christ Jesus, symbolizing their spiritual connection to the heavenly realm through knowing their identity in Christ Jesus and through consecration—embracing holiness, righteousness, and prayer.

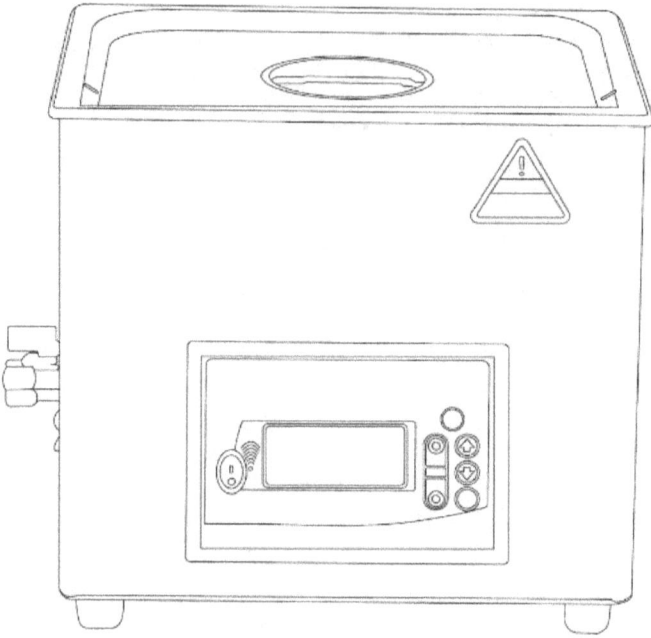

Figure 31: A sonicator is a device that uses sound waves to mix or break apart particles in a liquid.

Sonicator: Mixing with Holy Spirit

A sonicator is a scientific mixer used to blend substances together. It employs special sound waves to mix solutions and convert solids into liquids.

Biblical significance: The Fulfillment of God's Promise

Let's explore a significant event known as the Day of Pentecost, which recounts the fulfillment of a promise from God for the Twelve Disciples. As they gathered together, suddenly, a mighty sound, like a rushing wind, filled the entire place. Picture flames appearing and separating, descending upon each of them! What follows is even more remarkable: they were filled with the Holy Spirit and began speaking in diverse languages, aided by the Spirit (Acts 2:1-4).

Consider the Holy Spirit like the powerful waves in a sonicator. It descended upon the Disciples and the Children of God, resembling waves of sound, imparting the gift of speaking in unique languages and dwelling within them. This parallels how a sonicator blends substances; the Holy Spirit merged with them, becoming an essential part of their lives.

In 1 Corinthians 6:19-20, we find a powerful reflection: our bodies are described as temples of the Holy Spirit. This analogy encourages us to honor and care for our bodies just as we would a sacred place, recognizing that the Holy Spirit dwells within us. Since we were bought at a high price—the sacrifice of Jesus—it calls us to live in a way that respects and honors God through how we treat and care for our bodies.

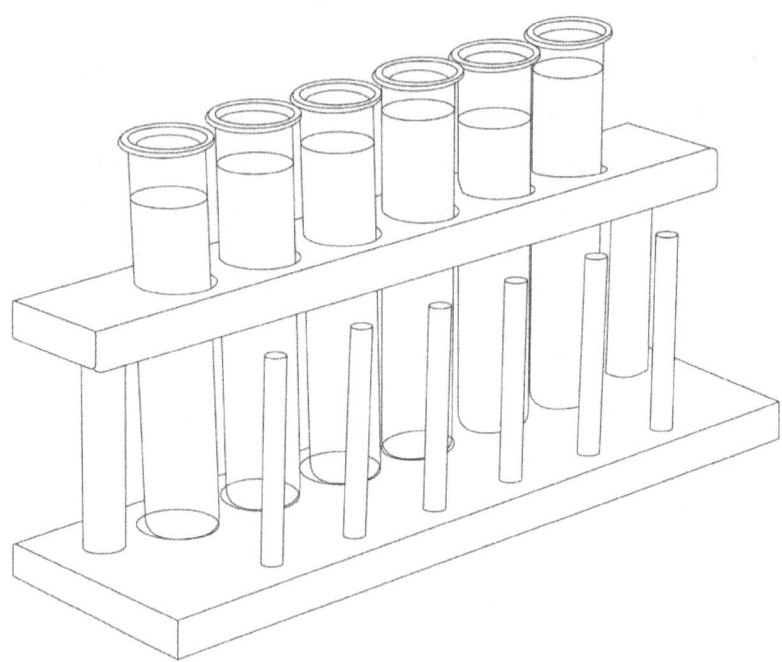

Figure 32: A test tube is a thin glass or plastic container used in laboratories for holding, mixing, or heating small amounts of liquids or solids.

Test Tube: Observing and Assessing Chemical Reactions

Imagine a small glass tube known as a test tube. It features an open end and a closed, rounded bottom. These tubes require a specialized holder called a test tube rack to remain upright. Scientists utilize test tubes for experiments, observing how various chemicals mix and react.

Biblical significance: Seek the Helping Hand of God

Life can be likened to a test tube experiment. We encounter challenges and difficult times that test our faith and resilience. Just as a scientist observes and assesses reactions in a test tube, God watches over us during our trials. These tough moments serve as tests for our hearts. God seeks to see whether we turn to Him for assistance or attempt to manage everything independently.

Dealing with challenges alone can feel like trying to touch a hot test tube with bare hands—exhausting and often resulting in failure. That's why we need God's help in every trial, especially when we feel weary. In Matthew 11:28, God is saying, *"Come to me, all of you who are tired and carrying heavy burdens, and I will give you rest."* Just as we use tongs to handle hot test tubes safely, we can lean on God during tough times.

Even when God seems silent during our trials, it's a reminder that He has already won the battle in the spiritual realm, and victory will manifest in the natural. Our faith is tested, and trusting in God, who never fails us, is crucial. Isaiah 43:19 beautifully illustrates that God can bring about new things, even in challenging situations. Trusting in Him is like discovering a way in the wilderness or finding rivers in the desert.

So, when life feels like a test tube experiment, remember to trust in the Lord, and He will guide you through.

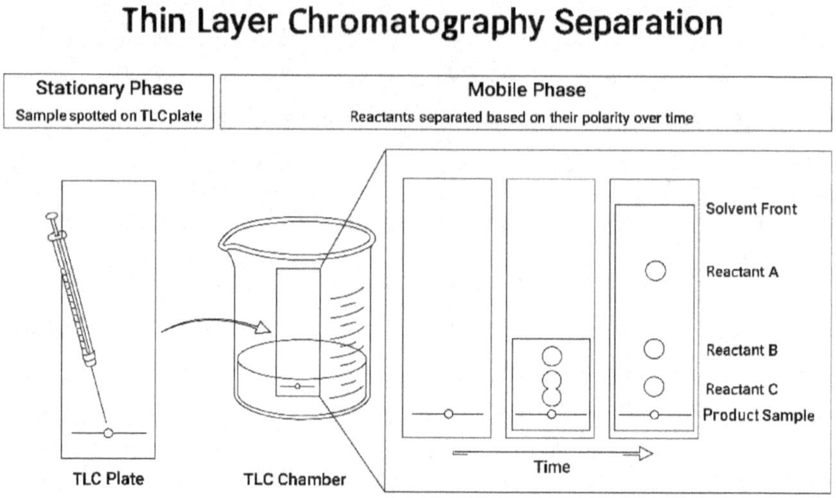
Thin Layer Chromatography Separation
Stationary Phase
Sample spotted on TLC plate
Mobile Phase
Reactants separated based on their polarity over time
Solvent Front
Reactant A
Reactant B
Reactant C
Product Sample
Time
TLC Plate
TLC Chamber

Figure 33: Thin layer chromatography (TLC) is a laboratory technique used to separate and identify compounds in a mixture by placing them on a thin layer of adsorbent material and allowing a solvent to move through it.

Thin Layer Chromatography (TLC): Purifies Mixtures

Thin Layer Chromatography (TLC) is a method that assists scientists in separating mixtures into their individual components. Scientists utilize TLC to monitor and assess the progress of a chemical reaction. A chemical reaction is a special transformation where the initial material undergoes changes, resulting in the creation of the final product. However, sometimes the reaction doesn't reach complete conversion, prompting the chemist to extend the experiment time to consume all the initial material. If traces of the starting material persist in the final product, the chemist may need to purify it using methods such as recrystallization, column chromatography, or other techniques.

Biblical significance: Partner with the Will of God

Just as a chemist grapple with ensuring a reaction reaches completion, people can face challenges in life when they lack knowledge of God's will. Without understanding God's plan, individuals might inadvertently align themselves with destructive forces. This lack of knowledge can result in cycles of defeat and bondage, perpetuating from one generation to the next, as covenants and agreements made in ignorance remain unbroken.

In Genesis 2:7, we learn that God formed man in His image, giving humans the ability to speak and create. Hebrews 11:3 highlights the creative power of faith, illustrating how worlds were framed by spoken words. This power of speech is further emphasized in Proverbs 18:21, which declares that the tongue has the ability to bring forth both life and death. When individuals accept sicknesses or hardships, making room for them in their lives, they inadvertently grant permission to destructive plans. However, aligning with God's word, proclaiming healing and deliverance, disrupts satan's schemes.

Hosea 4:6 emphasizes the significance of knowledge, stating that people perish due to lack of it. Maintaining a healthy lifestyle involves meditating on God's word day and night, as recommended in Joshua 1:8, which leads to prosperity and success. Romans 8:37 encourages believers to align their lives with God's word, making them more than conquerors. So, just as a chemist utilizes knowledge to purify a product, aligning with God's wisdom empowers individuals to break destructive cycles and live consecrated and victorious lives.

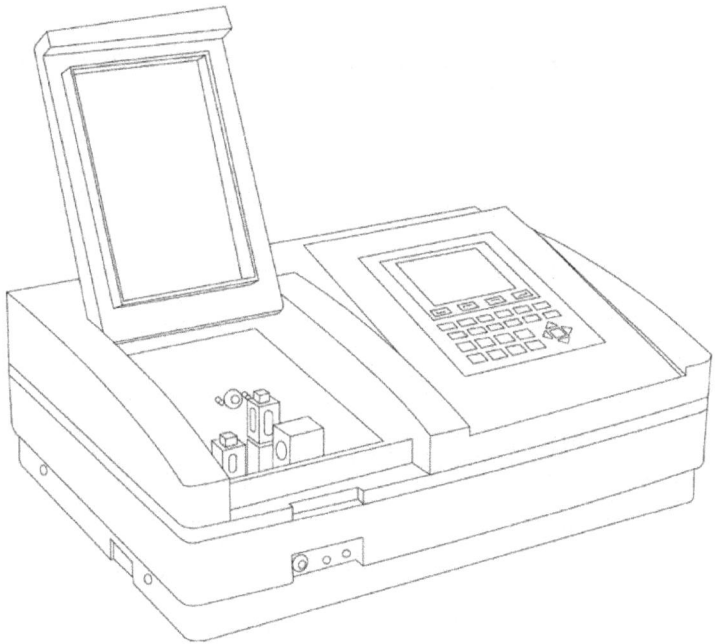

Figure 34: A UV-visible spectrophotometer is an analytical instrument that measures the absorbance or transmittance of ultraviolet and visible light by a sample to determine its concentration and chemical properties.

UV-Visible Spectrophotometer: Exploring Concentration

Imagine a remarkable tool known as a UV-Visible Spectrophotometer. It serves as a detective for scientists, aiding them in comprehending the concentration of various compounds in a solution. This tool operates by passing light through a sample solution and determining the amount of light absorbed by the solution. This technique is highly accurate and can provide scientists with precise information about the concentration of a specific compound in the solution.

Biblical significance: Immersing Ourselves in the Word of God (Jesus – The Living Water)

Let's draw a parallel to our daily lives, such as when we prepare a vitamin C drink. Picture this – you have a packet of vitamin C powder (the solute) that you want to dissolve in water (the solvent) to create your tasty beverage. The amount of powder you add to the water determines the concentration of your vitamin C drink.

Now, let's consider this as a lesson in solubility – the ability of a substance to dissolve in a solvent. In our scenario, the vitamin C powder dissolves in water. Similarly, in life, we encounter challenges or problems (the solute), and we need to dissolve them in the "solvent" (the living waters of God) of our determination, courage, and faith. Just as the UV-Visible Spectrophotometer helps determine concentration accurately, our focus on the word of God, the Bible, helps us overcome challenges.

So, when life presents challenges, view it as an opportunity to demonstrate your solubility, your ability to dissolve problems with discipline in meditating on biblical verses, prayer, and supplication. This lesson reminds us that, just like a well-prepared vitamin C drink, we can combine the right ingredients in life to create a solution that's strong, resilient, and full of goodness.

Figure 35: A vortex mixer is a lab tool that quickly stirs liquids in tubes or vials to mix them thoroughly.

Vortex Mixer – Seamless Blending

A Vortex mixer is a compact but powerful tool used in laboratories to blend liquids.

Biblical significance: The Mixing of God's Judgments

Just as a vortex mixer combines liquids in laboratories, the Bible illustrates God's use of a mixing process in His judgments. The phrase "the wages of sin is death" conveys that disobedience to God makes individuals vulnerable to their adversary, satan, as they may lose the protective covering of the Lord. In Jeremiah 50:35-42, God announces judgment upon nations, including Babylon, for their idolatry, ungodly political, social, and economic views. Revelation 18:6 vividly depicts the principle of reciprocation – paying back double for deeds, emphasizing a mixing of consequences.

A powerful example of God's use of mixing is found in Genesis 11:7. The Holy Trinity—God, Jesus, and the Holy Spirit—intervenes on earth, confusing and mixing up the language of the people who were united in building the Tower of Babel. This intervention prevents them from understanding one another's speech, deterring them from becoming prideful and working together for evil outcomes. The people's unity and determination to build a symbol of their greatness lead to this divine mixing, emphasizing the importance of humility and reliance on God rather than seeking self-glory and attempting to be like God.

Another example of God's righteous judgment is found in 1 Kings 3:8-15, when Solomon, newly ordained as King over Israel, asked God, *"Give your servant therefore an understanding mind to govern your people, that I may discern between good and evil, for who is able to govern this your great people?"* It pleased the Lord that Solomon had asked this. And God said to him, *"Because you have asked this, and have not asked for yourself long life or riches or the life of your enemies but have asked for yourself understanding to discern what is right, behold, I now do according to your word. Behold, I give you a wise and discerning mind, so that none like you has been before you and none like you shall arise after you. I give you also what you have not asked, both riches and honor, so that no other king shall compare with you, all your days. And if you will walk in my ways, keeping my statutes and my commandments, as your father David walked, then I will lengthen your days."*

So, just as a vortex mixer skillfully combines liquids in a laboratory, God employs a divine mixing process in His judgments to reward and teach lessons of humility, obedience, and reliance on Him.

Figure 36: Weighing paper is a thin, disposable paper used in laboratories to hold and weigh solid samples on a balance without contaminating the scale.

Weigh Paper: Carry's the Weight

A simple yet essential tool specifically designed for measuring the weight of solids on a balance is called a weigh paper.

Biblical significance: Just Weights and Precise Measurements

The concept of accurate measurements has deep roots in biblical principles, offering insights into the nature of God. In Proverbs 21:2, it's mentioned that while a person may think their ways are right, the Lord weighs the heart. This emphasizes the importance of accuracy and fairness in our thoughts and actions, aligning with God's desire for just weights and precise measurements. When individuals use accurate measurements, they reflect a commitment to honesty and integrity, fulfilling a godly request for fairness and righteousness.

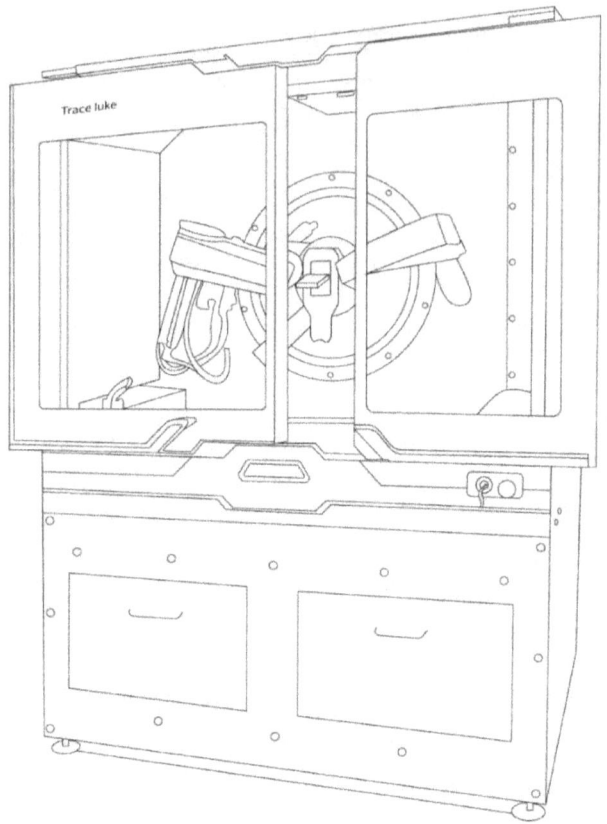

Figure 37: XRD (X-ray diffraction) is an analytical technique used to identify the crystalline structure of materials by measuring the pattern of X-rays scattered by the sample when irradiated with X-ray beams.

X-ray Diffraction (XRD) – The Molecular Arrangement of Atoms in Materials

XRD (X-ray diffraction) is a non-destructive analytical technique that uses high-energy X-rays to excite a single crystal in a material, whether crystalline or non-crystalline (amorphous). This method reveals the arrangement of atoms within the crystal. For example, when a grain or crystal of salt is analyzed using XRD, the technique unveils the specific arrangement of sodium and chloride atoms.

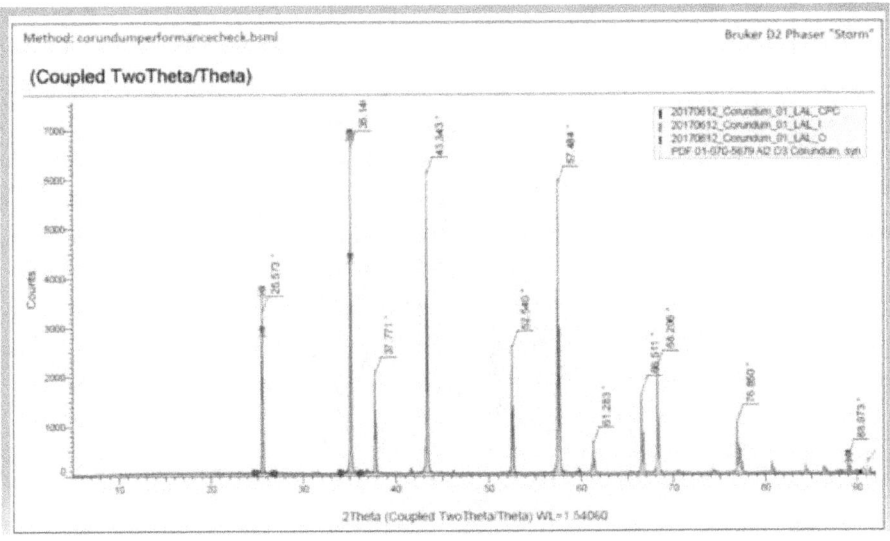

Figure 38: Spectral data of a corundum standard.

Biblical significance: Spiritual Arrangement in Jesus Christ

Just as X-ray diffraction (XRD) reveals the precise atomic arrangement within a crystal lattice, accepting Jesus Christ as our personal Lord and Savior initiates a profound spiritual realignment in God. This realignment transforms our lives as He forgives our sins, renews our souls, and grants us the promise of eternal life. The well-known verse, John 3:16, encapsulates this promise: *"For God so loved the world that he gave his one and only Son, that whoever believes in him shall not perish but have eternal life."*

In Jesus Christ, we discover our true purpose. Achievements in the world—such as degrees, wealth, and material possessions—

become insignificant in comparison to the eternal perspective. This is reflected in Ecclesiastes 1:12, which states: *"I have seen all the things that are done under the sun; all of them are meaningless, a chasing after the wind."* As ambassadors of Christ, our mission transcends earthly pursuits, centering on extending the kingdom of God here on earth, as emphasized in Luke 9:2: *After training His disciples, Jesus sends them to preach the kingdom of God."*

Figure 39: The fivefold ministry equips and unifies the body of Jesus Christ through Apostles, Prophets, Evangelists, Pastors, and Teachers to advance Elohim's Kingdom.

This mission is carried out through the fivefold ministry, which comprises the roles of Apostle, Prophet, Evangelist, Pastor, and Teacher. Each of these roles serves a unique function in equipping and building up the body of Christ:

- Apostles establish and oversee new church structures, laying the foundation for growth.
- Prophets discern and communicate God's will.
- Evangelists reach out to the lost, spreading the Gospel.
- Pastors nurture and guide the flock, offering care and counsel.
- Teachers interpret and impart scripture and doctrine.

These roles operate in harmony, empowered by the Holy Spirit, to cultivate a healthy and Spirit-led church. While some believers may manifest all five gifts, reflecting the ministry of Jesus, others may operate in specific capacities within the fivefold ministry, each contributing to the collective mission of the Church.

As children of the Living God, we gain a resilience that renders us indestructible by the enemy. This is achieved through dying to our fleshly desires, which often conflict with the laws and will of Elohim. Our commitment to God's laws and the assurance of eternal life make us spiritually unshakable. This truth is echoed in 2 Timothy 2:11: *"If we die with Him, we will also live with Him."* Jesus reaffirms this in John 11:25-26, declaring: *"I am the resurrection and the life. The one who believes in me will live, even though they die; and whoever lives by believing in me will never die."* Additionally, Romans 14:8 emphasizes this unbreakable bond: *"If we live, we live for the Lord; and if we die, we die for the Lord. So, whether we live or die, we belong to the Lord."*

Accepting Christ not only transforms our spiritual composition but also secures our eternal connection to the Father. This divine alignment empowers us to withstand spiritual challenges, making us resilient, steadfast, and eternally anchored in God's promises.

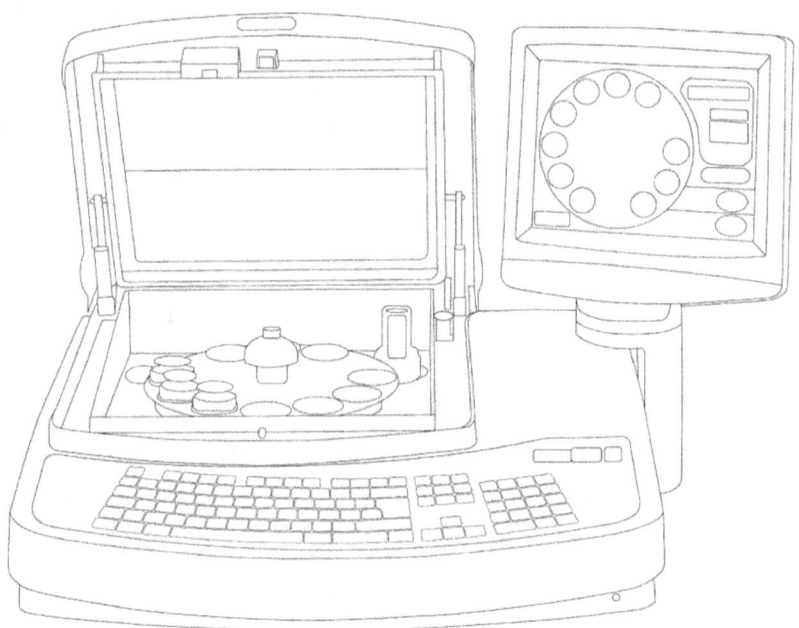

Figure 40: XRF (X-ray fluorescence) is an analytical technique used to determine the elemental composition of materials by measuring the fluorescent X-rays emitted from a sample when it is exposed to X-ray radiation.

X-ray Fluorescence (XRF) – Illumination of Materials Elemental Composition

XRF is a non-destructive analytical technique used to determine the elemental composition of solid, liquid, and powder materials.

Biblical significance: Understanding Our Elemental Composition

In the profound words of Maya Angelou, *"If you don't know where you've come from, you don't know where you're going."* This sentiment holds truth — understanding our origin provides a foundation for our identity and purpose. Without this knowledge, we become vulnerable to external influences that can shape our lives positively or negatively. However, when we recognize our belonging to Jesus Christ and find our identity in Him, we resist deception and live with a sense of royalty beyond human constraints.

To reach this awareness, we must close the door on sin, receiving God's forgiveness and acceptance. This journey parallels the promise land God assured our forefathers – Abraham, Isaac, Jacob, and Moses. To enter this land of abundance, we navigate through the refiner's fire, a spiritual test that breaks us down to our elemental form. In this refining process, the XRF metaphorically breaks us down, stripping away worldly characteristics and revealing the elements of God within us.

During this refining test, the Lord examines our hearts, evaluating whether we resemble Him. The testing uncovers what aspects of our lives align with godliness and what needs transformation. Crucial to passing the test is avoiding complaints and leaning on our own understanding. The Israelites' prolonged time in the desert serves as a cautionary tale – their complaints and lack of repentance delayed their entry into the promised land for forty years.

The testing period is designed to shape us into individuals resembling Christ Jesus, ensuring that when we receive blessings, we use them wisely for the Kingdom of God. Colossians 3:10 emphasizes putting on the new self, being renewed in knowledge in the image of our Creator, Elohim. Proverbs 17:3 illustrates the purpose of the refining process, likening it to the purification of silver and gold, with the Lord examining our hearts. Thus, the refiner's fire, symbolized by XRF, reveals our elemental composition, transforming us into reflections of God's image.

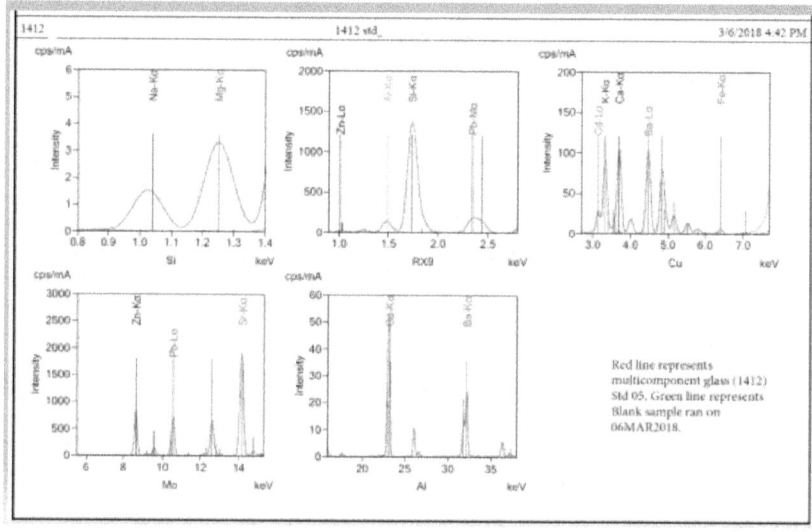

Figure 41: A spectrum of a sample analyzed by XRF.

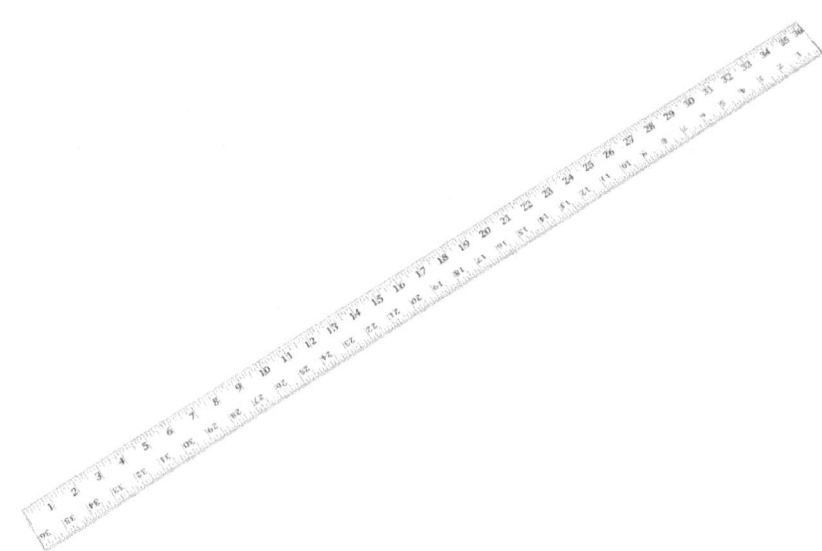

Figure 42: A yardstick is a measuring tool that is three feet long, typically marked with units of measurement, used for measuring length or distance.

Yard Stick: The Measurement of Objects

While a yardstick is a simple tool for measuring objects larger than a piece of paper, it can also serve as a metaphor for moral principles in our lives.

Just as a yardstick provides a fixed and consistent measure for physical objects, moral principles offer a stable and unwavering guide for our actions. These principles act as a moral yardstick, helping us measure our choices, behaviors, and decisions against a set standard.

Biblical significance: The Moral Yardstick of Life

In our journey through life, having a moral yardstick based on values such as honesty, integrity, compassion, and righteousness allows us to assess our actions. Galatians 5:22-23 expresses that when we are led by the Spirit of the living God, our attributes reflect the nine fruits of the Spirit: love, joy, peace, longsuffering, kindness, goodness, faithfulness, gentleness, and self-control. This moral measuring tool helps us navigate ethical challenges, ensuring that our conduct aligns with the principles of God we hold dear.

This sentiment is also illustrated in 2 Peter 1:5-8: *"For this very reason, make every effort to add to your faith goodness; and to goodness, knowledge; and to knowledge, self-control; and to self-control, perseverance; and to perseverance, godliness; and to godliness, mutual affection; and to mutual affection, love. For if you possess these qualities in increasing measure, they will keep you from being ineffective and unproductive in your knowledge of our Lord Jesus Christ."*

The yardstick, both in the physical and moral sense, represents a standard against which we can gauge ourselves. This moral philosophy is plainly stated in 1 Peter 1:14-16: *"But just as he who called you is holy, so be holy in all you do; for it is written: 'Be holy, because I am holy'."* Embracing a moral yardstick provides a foundation for personal growth, uprightness, and a meaningful ethical life. Just as we measure physical objects to ensure accuracy, measuring our actions against moral standards ensures the integrity and alignment of our character with the Lord.

Prayer for Heavenly Guidance and Provision

Our Abba, my full supplier in heaven, Your being is pure and perfect. There is none like You. Let Your Kingdom come to me on earth to continue to govern me, Your citizen, through the manifestation of Your holiness, righteousness, culture, and lifestyle. Abba, let Your original purpose and intent be accomplished in every area of my life here on earth.

My Lord and my God, change my world by destroying the evil systems in place on this physical planet. I pray for the same culture, standards, and lifestyle of the kingdom of heaven to come to me here on earth.

Lord, let there be order, peace, perfection, joy, love, worship, and songs in my life. Abba, supply me and the community of citizens with the demands and provisions for each day. God, please provide all things necessary for life: friendships, divine helpers, profitable networks, good health, and a sound mind.

Father, forgive us, the community, of our sins that led us astray from Your purpose for our lives. Father, we release and forgive those who infringed on our domain and purpose. Abba, help us to follow Your Spirit and discernment and stop testing Your boundaries.

Lord, keep us from the one who will distract us from Your purpose and will for our lives and for the earth. For I am not of my own kingdom because I am under Your governance and Yours alone.

Thank You, Father, for Your power and abilities to provide me with the kingdom of heaven lifestyle and culture. Your glory shall impact and be imprinted on me. I agree with You, God, and submit to Your purpose. Amen.

The Power of Fasting

A Comprehensive Guide to Fasting for Spiritual Growth and Purpose

Fasting is a profound spiritual discipline that goes beyond abstaining from food; it is a purposeful practice designed to deepen our connection with God, promote spiritual growth, and align us with His divine will. This guide explores the different types of fasting, essential principles to follow, and practical steps to maximize the impact of your fasting journey.

Through fasting, we create space to hear God's voice more clearly, seek His guidance, and nurture a spirit of humility. By choosing a fast that fits your spiritual goals and capacity—whether a dry fast, water fast, Daniel fast, or intermittent fast—you can intentionally engage in worship, prayer, and Bible study. Fasting quiets distractions, elevates faith, and draws us nearer to God's presence and purpose, serving as a vital tool for transformation and spiritual renewal.

Reasons for Fasting

Here are several key reasons why fasting can be transformative for our faith journey:

1. **Drawing Closer to God**
   One of the primary reasons for fasting is to deepen our connection with God. Through fasting, we redirect our

focus from daily distractions to worship, prayer, and meditation on God's word, the Bible. This intentional pursuit helps us become more attuned to God's presence, fostering a closer and more personal relationship with Him.

2. **Seeking God's Guidance**

Fasting also aids us in seeking God's direction, especially when facing significant decisions or challenges. By setting aside time for fasting, we create space for God to speak to us, offering clarity and guidance as we navigate life's complexities. It's a way to open our hearts and minds to the Holy Spirit's leading, allowing God to guide us toward the right path.

3. **Humbling Ourselves Before God**

Fasting humbles us before the Lord, serving as a reminder of our complete dependence on Him. This act of humility is a way of expressing our need for God and recognizing our inability to succeed on our own. It often leads to spiritual breakthroughs, empowering us to overcome obstacles that may hinder our growth. In this humbled state, we become more receptive to God's grace, which brings renewal and strength.

4. **Interceding for Others**

Fasting can also be a means of intercession, allowing us to pray fervently on behalf of others. As we fast, we bring their needs before God, lifting up situations that require His intervention. This focused intercession builds compassion in us, and as we commit to praying for others, God's love moves through us, deepening our empathy and solidarity with those in need.

5. **Repentance and Renewal**

Fasting allows us to repent, seek God's forgiveness, and renew our commitment to Him. 2 Chronicles 7:14 states, *"If My people who are called by My name will humble themselves, and pray and seek My face, and turn from their wicked ways, then I will forgive their sin and heal their land."* Humbling ourselves in fasting demonstrates our willingness to deny the desires of our flesh, coming to God in surrender and meekness. It is a time of self-reflection, where we examine our hearts, recognize areas for growth, and make necessary changes to realign with God's will.

6. **Obedience and Spiritual Strength**

Following the biblical examples of fasting is an act of obedience that strengthens our faith. This discipline aligns us with God's commandments, builds resilience, and reinforces our spiritual resolve. Denying ourselves physical comforts not only heightens our spiritual awareness but also cultivates a deeper reverence for God, as we resist indulgence and practice self-control. Fasting restores humility, allowing us to acknowledge God as our ultimate source of strength.

7. **Serving Others and Showing Compassion**

In Isaiah 58, we are reminded of the importance of caring for the needy during a fast, which calls us to show compassion and address the needs of others. Fasting is not solely about abstaining from food or drink but also about aligning our hearts with God's love for the poor. Proverbs 19:17 says, *"Whoever is kind to the poor lends to the Lord, and he will reward them for what they have done."* This compassion fosters a greater awareness of the struggles others face, helping us become instruments of God's love.

8. **Preparation for Ministry and Service**

    Lastly, fasting equips us for ministry and service by aligning our hearts with God's will and purpose. This time of devotion strengthens us to serve God effectively and to fulfill the unique calling He has placed on our lives.

Types of Fasts

Fasting is a powerful spiritual practice that allows us to humble ourselves, draw closer to God, and seek His guidance. While the method of fasting varies, the central goal remains the same: to focus on worship, prayer, and Bible study, deepening our connection with God. Below are key principles for a successful fast and descriptions of various types of fasting to consider:

Key Principles for Fasting

1. **Focus on Worship and Prayer**

    The primary purpose of fasting is to grow closer to God. Engage in worship, prayer, and Scripture reading multiple times daily.

2. **Keep Your Fast Private**

    Fasting is an intimate experience between you and God. As

Jesus taught in Matthew 6:16-18, avoid broadcasting your fast to seek others' approval. Instead, focus on God, who sees what is done in secret and will reward your faithfulness.

3. **Reflect on Isaiah 58**

   Isaiah 58 outlines God's expectations for fasting, emphasizing that true fasting includes helping those in need and acting with compassion and humility.

4. **Practice Generosity**

   Give to the poor and needy during your fast (Isaiah 58). This selflessness aligns your heart with God's, cultivating compassion and empathy.

5. **Set Clear Intentions**

   Journal your intentions, listing three primary prayer points or specific areas where you seek God's guidance.

6. **Seek the Holy Spirit's Guidance**

   Ask the Holy Spirit to direct you in choosing the type of fast and its duration. Fasting is a personal agreement with the Lord, so tailor it according to His leading.

7. **Avoid Focusing on Food**

   Fasting is not about food abstinence alone. Philippians 3:19 warns against allowing earthly desires, like food, to become a distraction. Instead, focus on prayer and spiritual growth.

8. **Combine Fasting with Prayer**

   Fasting without prayer is merely a diet. Pair your fast with consistent, intentional prayer to give it spiritual purpose.

9. **Persevere Through Challenges**

   Mistakes happen. If you stumble during your fast, don't quit. Ask for forgiveness, refocus, and continue. Over time, fasting will become more natural and fulfilling.

10. **Consider an Accountability Partner**

   An accountability partner can provide support and encouragement. Regular fasting can strengthen your relationship with God and maintain a spiritual focus as ambassadors of Christ in the world.

11. **Stay Positive and Trust in God**

   Avoid dwelling on your struggles. Speaking negatively can undermine your trust in God's ability to work in your situation. Trusting God fully during your fast demonstrates your belief in His sovereignty.

12. **Consume Uplifting Content**

   During your fast, focus on consuming content that strengthens your faith, such as worship music and sermons that honor Jesus. Avoid distractions like social media and other content that doesn't uplift your spirit. This dedicated time of fasting and prayer is an opportunity to train yourself to recognize God's voice—whether through Scripture, in prayer, audibly, or within your heart.

13. **Journal Your Experience**

   Document your reflections and insights throughout the fast to capture spiritual growth and answered prayers.

14. See God's Word as Your Nourishment

The Bible is spiritual food. As you lean on God's Word, hunger pains may subside, reminding you that spiritual sustenance is found in God alone. Matthew 4:4 says, *"But He answered and said, "It is written, Man shall not live by bread alone, but by every word that proceeds from the mouth of God."*"

15. Set a Schedule

Create a daily schedule for prayer, worship, and Bible reading to stay consistent and intentional throughout your fast.

Recommended Types of Fasting

1. Dry Fasting

A complete fast from food and water for a specific period. This is an intensive form of fasting, seen in the Bible when Esther called a three-day fast (Esther 4:16) and Moses fasted for forty days on Mount Sinai (Exodus 34:28).

2. Water Fasting

Abstaining from food but drinking only water. Jesus's forty-day fast in the wilderness is often seen as a water fast, as he experienced hunger but no mention is made of thirst (Matthew 4:2).

3. Daniel Fasting

Consuming only fruits, vegetables, and water for a designated time. Daniel practiced this when he requested to be tested on a diet of vegetables and water (Daniel 1:12).

4. **Intermittent Fasting**

   Alternating between periods of eating and fasting within a structured time frame, such as fasting from 6:00 a.m. to 6:00 p.m. This approach allows a balance between fasting and nutritional needs and can be adapted for specific hours in the day.

Example Schedule for Fasting

This schedule provides a structured approach to ensure consistency in worship, prayer, and Bible reading throughout the fast. You may adjust the times to fit your personal routine.

Days	Activities	Times
Day 1	Worship, Pray, Bible Reading	12 a.m., 12 p.m., 6 p.m.
Day 2	Worship, Pray, Bible Reading	9 p.m., 3 a.m., 12 p.m.
Day 3	Worship, Pray, Bible Reading	6 p.m., 12 a.m., 9 a.m.

This schedule is a guide to help you maintain consistency. Adjust activities and times as needed, focusing on deepening your connection with God.

In Summary:

Choose a type of fast that aligns with your spiritual goals and capacity—whether a dry fast, water fast, Daniel fast, or intermittent fast. Approach fasting with a spirit of humility, and set clear intentions for your journey by dedicating time daily to worship, prayer, and Bible study. Fasting provides a unique opportunity to step back from daily distractions, refocus your heart on God, and allow your faith to deepen.

By understanding and embracing the reasons for fasting, it becomes a transformative practice that fosters spiritual growth, humility, and a closer connection with God. Fasting reminds us of our dependence on Him, opening our hearts to His guidance and wisdom. As we draw closer to His presence, we gain clarity, compassion, and a renewed sense of purpose, equipping us to live a life more fully aligned with God's will.

Conclusion

"The Chemistry of God" unveils a unique perspective, showing how the laboratory of life is infused with profound Biblical significance and sacred insights. By drawing similarities between science and devotion, we've uncovered the hidden wisdom that surrounds us. As you continue your journey through life's experiments, may you find inspiration and guidance in the connections between the laboratory and the Holy Trinity of God.

Glossary

Actinic – A word that describes materials that can react chemically when exposed to light, especially ultraviolet (UV) light.

Active Pharmaceutical Ingredient (API)– It's the component in a medication that produces the intended therapeutic effect or action in the body.

Analyte – An analyte is a substance or chemical component that is being measured or analyzed in a sample during a laboratory test.

Analytical Balance – An analytical balance is a highly sensitive laboratory scale used to measure the mass of substances with great precision, often down to the milligram or microgram level.

Analytical Testing – Analytical testing refers to a set of laboratory procedures used to identify and quantify the chemical composition, structure, and properties of a sample to ensure quality and compliance with standards.

Armor of God – Protective gear worn by the children of God in the spirit realm to guard them from evil attacks.

Beakers - Beakers are cylindrical glass or plastic containers used in laboratories for mixing, stirring, and heating liquids, often featuring a spout for easy pouring.

Believe – To believe means to think something is true or to have faith in something without always needing to see it. It's like trusting in the goodness of God even if you can't touch or see Him.

Centrifuge – A laboratory device that spins samples at high speeds to separate substances based on their density, allowing for the isolation of particles or liquids within a mixture.

Chemicals – Distinct substances composed of specific atoms or molecules, used in various applications such as research, industry, and everyday life, each with defined properties and reactions.

Composition – Composition refers to the makeup or arrangement of different elements or compounds in a substance, determining its characteristics and properties.

Compounds – Compounds are substances made of two or more elements that are chemically combined in fixed amounts, creating materials with unique properties different from the individual elements.

Consequences – The results or outcomes that happen because of the choices we make. If you decide to do your homework, the consequence might be getting good grades. If you forget your umbrella on a rainy day, the consequence could be getting wet. It's like the "what happens next" part of our actions.

Consistently – Doing something the same way over and over again. Consistency is being reliable and doing things regularly!

Counsel – Devine advice from the Spirit of the Lord to navigate life. It's like talking to a wise friend when you have questions or need guidance. They give you good ideas to make good choices! The Spirit of counsel allows you to consistently receive advice from a reliable and good source, greater than human beings.

Desiccator – A special container that helps keep materials dry by removing moisture from them.

Dissolution Testing – A laboratory procedure used to measure how quickly and completely a drug dissolves in a liquid, helping to evaluate its release characteristics and absorption in the body.

Elements – Elements are pure substances that cannot be broken down into simpler substances and consist of only one type of atom, each defined by its atomic number and unique properties.

Emitted – When something is sent out or given off, like a smell from flowers or when a flashlight sends out a beam of light.

Erlenmeyer Flask – A container used in laboratories for mixing, heating, and holding liquids, featuring a narrow neck that minimizes evaporation and allows for easy swirling without spilling.

False gods – False gods are idols or deities (fallen angels) worshipped instead of the true God (Elohim), often representing human-made beliefs, people or objects.

Fear of the Lord – "Fear of the Lord" is a phrase from the Bible that doesn't mean being scared of God. Instead, it's about having deep respect and awe for the goodness, love, and wisdom of God. This deep admiration for God keeps us from breaking his laws and disappointing him.

Filter paper – A porous paper used in laboratories to separate solids from liquids by allowing the liquid to pass through while retaining the solid particles.

Flammable Cabinet – A specialized storage unit designed to safely contain and store flammable liquids and materials, minimizing the risk of fire hazards in laboratories or industrial settings.

Fourier Transform Infrared Spectroscopy (FTIR) – A technique used to identify materials by measuring how they absorb infrared light, revealing information about their chemical bonds and structures.

Frequency – Frequency refers to the number of times a repeating event occurs in a given time period, typically measured in hertz (Hz), which represents cycles per second.

Gas Chromatography Mass Spectrometry (GC-MS) – An analytical technique used to separate, identify, and quantify compounds in a sample by first vaporizing the sample in gas chromatography and then analyzing the resulting fragments with mass spectrometry.

Gas Gradient – In a GC-MS or HP-LC system, a gas gradient refers to a controlled change in the pressure or concentration of a gas used to transport gaseous analytes from a sample mixture through the column to the detector. The flow rate of the gas is gradually adjusted over time to enhance the separation and detection of substances in the sample. This gradual change improves peak resolution, making it easier to identify and quantify the components of the mixture.

Graduated Cylinder – A tall, narrow container marked with measurement lines used to accurately measure the volume of liquids.

Holy Spirit – The Holy Spirit is like a special friend from God. He's not something you can see, but is always there to help, comfort, and guide you. It's like having a gentle and wise companion inside your heart, reminding you to be kind and make good choices.

Honest – Being honest means telling the truth and not pretending. It's about being fair and trustworthy, so others can rely on you and know you're doing the right thing.

Hot Plate – A hot plate in chemistry is a flat, electrically powered device used to heat substances in a laboratory setting.

Hygroscopic – A word that means something likes to absorb or take in moisture from the air.

Identification – Identification is the process of determining the identity of substances or compounds using techniques such as chromatography and spectroscopy to gather information about their properties and behavior.

Incubator - A controlled device used to maintain specific temperatures and conditions for growing and storing microbial cultures or conducting reactions.

Integrity - Integrity means always doing what you know is right, even when no one is watching. It's like having a strong moral compass that guides you to make good choices and be a person others can trust.

Intensity Counts – Intensity counts refer to the measured strength or amount of a signal, indicating the abundance or concentration of a specific substance in analytical techniques like spectroscopy or mass spectrometry.

Ionizes – Substances can ionize when they gain or lose electrons, becoming positively or negatively charged particles called ions.

Jig – In GC-MS (Gas Chromatography-Mass Spectrometry), a jig tool is a specialized device used to securely install the chromatography column into the inlet and transfer line, ensuring proper alignment and stability for accurate analysis.

Karl Fischer Titrator – A Karl Fischer titrator is an instrument used in chemistry to precisely measure the water content in a sample by performing a titration.

Knowledge – A spirit of the Lord who reveals information that are not easily attained through natural means.

Laboratory – A controlled environment equipped for conducting scientific research, experiments, and measurements, typically involving various tools, equipment, and materials.

Liquid Chromatography Mass Spectrometry (LC-MS) is an analytical technique used to separate, identify, and quantify components in a mixture by passing a liquid sample through a column packed with solid adsorbent material under high pressure.

Living water – The "living water" is a symbolic term used to represent Jesus Christ who is pure and brings life and fulfillment. Jesus Christ is a gift from God that satisfies our deepest needs and helps us grow spiritually, just as water is essential for life on Earth.

Mass -to- Charge Ratio (m/z) – It's the ratio of an ion's mass to its charge, which means it tells us how heavy the ion is compared to how much electric charge it carries.

Materials – Materials are the substances or ingredients that make up everything around us, from medicines to buildings. They include various types like wood, metal, plastic, and more, each with unique properties that make them suitable for specific uses.

Matter – Matter is anything that has mass and occupies space, consisting of atoms and molecules, and can exist in different states, such as solid, liquid, or gas.

Measurement – Measurement is the process of determining the size, quantity, or degree of something by using tools or techniques to obtain accurate and precise values.

Melting Point - A melting point is the temperature at which a solid substance turns into a liquid.

Meniscus - The meniscus is like a curved water shape you see in a glass or a container, showing the level of a liquid, such as water or oil, where it's higher at the edges and lower in the middle.

Metaphor – A metaphor is a figure of speech that describes one thing as if it were another to highlight a similarity, like saying "His smile was bright like the sun" to show how big and happy his smile was.

Might – A strong and powerful spiritual being of the Lord who strengthens us to start and complete divine assignments given to us from God.

Mixtures – Mixtures are combinations of two or more substances that are physically blended but not chemically bonded, allowing each substance to retain its own properties.

Nitrogen Tank – The nitrogen tank stores nitrogen, an inert gas commonly used to create a controlled, oxygen-free environment.

Non-destructive – A way scientists can figure out what things are made of without breaking or harming them.

Oven – A controlled heating device used to dry, decompose, or heat substances at precise temperatures.

Personal Protective Equipment (PPE) is specialized gear used in science labs and various work environments to protect workers from hazardous chemicals, biological agents, physical hazards, and other potential dangers. PPE may include gloves, lab coats, goggles, face shields, respirators, and more, each designed to safeguard against specific risks and ensure a safe working environment.

Praying in Tongues – Praying in tongues is a special way some people talk to God using words that might sound different and not like the regular words we use every day. It's like having a secret language between them and God, expressing their feelings and thoughts in a unique and heartfelt way.

Qualitative Data – This type of data describes qualities or characteristics. It is non-numerical and often involves categories or descriptions, such as colors, names, or opinions. For example, responses to a survey about favorite ice cream flavors would be qualitative. Words to describe qualitative data are descriptive, detailed, open-minded, subjective, interpretive, contextual, insightful, and inductive.

Quantitative Data – This type of data involves numbers and measurements. It can be counted or measured and is often used for statistical analysis. For example, the number of ice cream cones sold in a day or the average temperature in a week would be quantitative data. Words to describe quantitative data are precise, objective, fixed, and quantifiable.

Recrystallization – The recrystallization method is a technique used to purify solid compounds by dissolving them in a hot solvent and then allowing them to slowly crystallize as the solution cools.

Resources – Resources are materials or assets used to meet needs or achieve goals, like natural resources, skills, or money.

Retention Time – It's the time it takes for each component of a mixture to travel through the chromatography column and reach the detector.

Revelation – Learning something important or getting a message from God that helps you understand things better, like finding a hidden treasure of wisdom or a special secret.

Reverence – A deep respect and awe for something significant or sacred, such as feeling amazed by the beauty and power of the Lord.

Righteousness – Righteousness means doing what is fair and good, standing up for what you believe is right, and treating others with kindness and fairness.

Rotary Evaporator Process – A process in chemistry used to remove solvents from a sample by heating it in a rotating flask under reduced pressure. This technique efficiently evaporates the solvent, leaving behind the desired compound.

Scientist - A scientist is someone who explores and learns about the world by asking questions, conducting experiments, and making discoveries to understand how things work.

Scripture – Special writings in the Bible that are holy and sacred. These writings often contain stories, teachings, and important lessons that are meaningful to people. It's like a guidebook that helps people understand how to live good and kind lives.

Solubility – It's how much of something, like sugar or salt, can be mixed with water or other liquids until it won't disappear anymore.

Solute – A solute is a gas, liquid, or solid that dissolves or disappears in a solvent like sugar in water.

Solution – A mixture where one thing (like sugar) disappears into another thing (like water) to make something new (like sweet water). It's when different parts mix together so well you can't see them separately anymore!

Solvent – A liquid that helps substances dissolve like water that makes sugar or salt disappear when you mix it, creating a solution.

Sonicator - A device that uses ultrasonic sound waves to agitate and break apart particles or samples, aiding in processes like mixing, extraction, or homogenization.

Spectrum – A spectrum is a graph that scientists use to analyze and understand the different components or characteristics of a substance, visually representing data from a test to help interpret results.

Substance – A type of matter with a specific composition and distinct properties, such as water, salt, or gold.

Test Tube – A test tube is a thin glass or plastic container used in laboratories for holding, mixing, or heating small amounts of liquids or solids.

Thin Layer Chromatography (TLC) – Thin layer chromatography is a laboratory technique used to separate and identify compounds in a mixture by placing them on a thin layer of adsorbent material and allowing a solvent to move through it.

Transform – A deep, inner change in a person's heart and spirit, such as becoming kinder and a better version of oneself through God's love or teachings. Transformation involves growing into a more loving person and becoming a new, improved version. This process is achieved by renewing our minds with the teachings of the Bible.

Trust – Having confidence in someone or something, believing they will fulfill their promises based on their past actions. Trust acts like a strong bridge between God and people, strengthening and solidifying relationships.

Understanding – A spiritual being of the Lord who assists you in connecting ideas like putting puzzle pieces together to see the bigger picture.

UV-Visible Spectrophotometer (UV-Vis) – A UV-visible spectrophotometer is an analytical instrument that measures the absorbance or transmittance of ultraviolet and visible light by a sample to determine its concentration and chemical properties.

Vortex Mixer – A vortex mixer is a lab tool that quickly stirs liquids in tubes to mix them thoroughly.

Wisdom – A spiritual being of the Lord who guides us in making good decisions. She instructs you in how to proceed by providing the necessary steps for success in life.

X-ray Diffraction (XRD) – A analytical technique used to identify the crystalline structure of materials by measuring the pattern of X-rays scattered by the sample when irradiated with X-ray beams.

X-ray Fluorescence (XRF) – An analytical technique used to determine the elemental composition of materials by measuring the fluorescent X-rays emitted from a sample when it is exposed to X-ray radiation.

Yardstick – A yardstick is a measuring tool that is three feet long, typically marked with units of measurement, used for measuring length or distance.

For the Children of Elohim

www.ingramcontent.com/pod-product-compliance
Lightning Source LLC
Chambersburg PA
CBHW070427290526
45791CB00005B/1864